# RURAL COOPERATIVES IN SOCIALIST UTOPIA

# RURAL COOPERATIVES IN SOCIALIST UTOPIA

## Thirty Years of Moshav Development in Israel

Edited by

Moshe Schwartz,
Susan Lees, and Gideon M. Kressel

PRAEGER

Westport, Connecticut
London

**Library of Congress Cataloging-in-Publication Data**

Rural cooperatives in socialist utopia : thirty years of moshav development in Israel / edited by Moshe Schwartz, Susan Lees and Gideon M. Kressel.
p. cm.
Includes bibliographical references and index.
ISBN 0-275-95309-2 (alk. paper)
1. Moshavism. 2. Agriculture, Cooperative—Israel. I. Schwartz, Moshe. II. Lees, Susan H. III. Kressel, Gideon M.
HD1491.I7R87 1995
307.72'095694—dc20 95-11276

British Library Cataloguing in Publication Data is available.



Library of Congress Catalog Card Number: 95-11276
ISBN: 0-275-95309-2

First published in 1995

Praeger Publishers, 88 Post Road West, Westport, CT 06881
An imprint of Greenwood Publishing Group, Inc.

Printed in the United States of America

The paper used in this book complies with the Permanent Paper Standard issued by the National Information Standards Organization (Z39.48-1984).

10 9 8 7 6 5 4 3 2 1

# Contents

# Tables and Figures

**TABLES**

## FIGURES

# Preface

In the 1950s, thousands of Jewish immigrants to Israel found themselves subject to a Zionist/Socialist project. They were settled in small rural cooperatives called *moshavim.* Here, they were provided with homes, land, agricultural instruction, and indoctrination into the distinctive pattern of cooperative economy that was to govern their lives. They learned their lessons with skill and speed, and many moshavim developed rapidly into prosperous and productive agricultural communities—so much so that other nations looked to the moshav model to develop their own rural economies.

For many Israelis, the moshav, though less radical than the more famous kibbutz, symbolically represented the fulfillment of a Zionist/Socialist dream—the creation of a new Jew and a new Jewish society based on principles of cooperation and dedicated physical labor. But in the thirty or more years that have passed since the hurried establishment of the moshavim of the 1950s, there have been many changes in both the moshavim themselves, as they developed and matured and confronted a variety of intense political, economic, demographic, and ecological pressures, and the larger Israeli society.

In order to consider these changes and their impact on the moshavim, in 1990 Moshe Schwartz and Gideon Kressel of the Ben-Gurion University of the Negev convened a special symposium comprised of scholars with expertise on this subject. Some of these scholars had studied specific moshavim in the 1950s and 1960s. Some had maintained a continuing relationship with "their" moshavim for decades. Others had studied the institutions that had shaped moshav life from outside and had traced the history of changing relationships between moshavim and these institutions. Together, these scholars decided to revisit the sites and institutions of their earlier research and to reconvene in 1991 to report to one another the results of their restudies.

This book, edited by Moshe Schwartz, Susan Lees, and Gideon Kressel, contains the outcome of the second meeting. Some chapters bring us up-to-date on changes taking place in individual moshavim two or three decades after the initial study. Some discuss the history of institutional change, while others provide an insider's view of changes in moshav life. Finally, some look at the export of the moshav model to other countries to evaluate its success elsewhere. Given the collapse of socialism in Eastern Europe and the former Soviet Union and the crisis in which the Israeli socialist rural institutions, the moshav and the kibbutz, find themselves today, we believe this collection of original articles is particularly timely.

After an introductory overview of the moshav as an institution by Susan Lees, Chapter 1 by Moshe Schwartz provides a history of moshav research since the 1950s, attempting to analyze the part played in it by various government institutions, academic disciplines, and schools of thought.

Chapter 2 by Neal Sherman and Moshe Schwartz also discusses the Settlement Department. The authors analyze the impact of the financial support the Settlement Department provided in the early 1980s to the post-1967 moshavim of the Jordan Valley upon the conduct of cooperative economic affairs. They show that public funding gravely undermined cooperative discipline, leading to the accumulation of enormous debts, the likes of which would be accrued by the whole settlement sector a few years later.

In Chapter 3, Shlomo Fox, an economist, deals with the evolution of the Settlement Department of the Jewish Agency and shows that the demise of the Department parallels a gradual though radical change in its relationship to the rural sector, as well as in the types of new rural settlements being set up in Israel, from agricultural settlements to commuter communities.

In Chapter 4, Susan Lees discusses the problem of efficiency in water distribution and use. She shows how the transformation of farming from a way of life (cooperative) to a type of business (capitalist) has prepared the ground for a substantial policy shift in the treatment of water as a factor of production.

In Chapter 5, Esther Schely-Newman discusses her return to the moshav where she grew up, which was set up in the early 1950s for immigrants from Tunisia. Schely-Newman gathered the stories told by the early settlers about their early days in the settlement. In their stories, a mythical structure emerges, legitimizing settlement and imbuing the settlers with pride. The settlers are thus recast as sucessful rather than reluctant pioneers, as well as full-fledged members of Israeli society.

Naomi Nevo also went back to her old moshav, Nahalal, which was established as the country's first moshav in 1921. In Chapter 6, she describes patterns of organization and relations which have remained as they were, even though much has changed in the environment.

In Chapter 7, Harvey Goldberg returns to the site of his 1960s ethnography of a community of Libyan immigrants. He updates the tale of adjustment undergone by the villagers since that time but takes note of important continuities in cultural practices and ethnic identity.

Moshe Shokeid has maintained contact with the settlers of a moshav set up in the 1950s in the Negev for immigrants from the Atlas Mountains. He had studied them in the early 1960s and had met their young children. In Chapter 8, he discusses the careers of the second generation in this moshav and nearby moshavim. The chapter shows that the founders are not always interested in keeping their offspring in the moshav and explains why.

Leonard Mars, a social anthropologist of the Manchesterian persuasion, returned after twenty years to the moshav he studied in the early 1970s. It was set up in the early 1950s by immigrants from Eastern Europe, many of whom survived the Holocaust. In Chapter 9, Mars shows how his moshav has lost its agricultural and cooperative characteristics and how it has come increasingly to share the characteristics of commuter communities, supporting a few workshops and small industries.

In Chapter 10, Gideon Kressel discusses the generation gap. In this chapter, he discusses the attitudes toward manual labor in a moshav set up by Latin American immigrants a short distance from Tel Aviv. The children of this moshav have lost the respect that their parents had toward agricultural labor. Kressel explains the impact of cheap labor from Gaza in the process of attitude change.

In Chapter 11, David Sivan presents an exception, a community which is thriving on agriculture and remains committed to cooperation. He attributes the community's special characteristics to its religious affiliation to the Karaite sect.

Yair Levi, a political scientist and cooperative expert, discusses the attempts to export the Israeli moshav model to Third World countries. In Chapter 12, he analyzes the motives and beliefs that underlie these attempts. He reviews a few cases in Asia, Africa, and Latin America and revisits the issues in the light of current development, views, and trends.

In Chapter 13, Moshe Schwartz reviews the transformations which Israeli moshavim have witnessed from the mid-1980s to the present. He analyzes the processes of decooperativization, changing occupational structure, and suburbanization and shows how these transformations are related to the changes which have occurred in the organizational and social environments in which the moshavim operate.

Finally, Chapter 14 by Raanan Weitz places these changes in a larger theoretical context. Weitz examines changing relationships between local communities and the state as economic development alters the environment for economic cooperation.

The editors acknowledge the support of the Jacob Blaustein Institute for Desert Research, the Ben-Gurion University of the Negev, and Hunter College, as well as the assistance of graduate student assistants Molly Doane and Kee Howe Yong.

# The Moshav: An Introduction

*Susan Lees*

## EXPERIMENT OR WAY OF LIFE?

Israel's most famous rural institutions, the kibbutz (plural, kibbutzim: collective settlement) and the moshav (plural, moshavim: small holder cooperative settlement), have been referred to almost uniformly in both scholarly and popular literature as experiments. In fact, this is a mistake and a misapprehension. They are, like many other social institutions, the combined result of deliberate design and accommodation to circumstances through time. But neither the design nor the accommodation were made as experiments; rather, they have been largely self-conscious efforts by people to work out ways to live and survive that (1) express some ideals, (2) respond to problems, opportunities, and constraints in the larger social and natural environment, and (3) serve the needs of both individual human beings and the society in which they exist. Their innovativeness, as well as their deliberateness, may make these efforts and their results appear to be experimental; but, in fact, they do not resemble true experiments at all. Their intent was not to satisfy intellectual curiosity but to create new ways of life and to cope with difficult circumstances.

There are many reasons, however, to want to know what has happened to the kibbutz and the moshav over time, including intellectual curiosity among them. These institutions attracted worldwide attention because many of the problems they addressed appeared similar to problems elsewhere—problems of how to satisfy social ideals of equality and justice in a rural setting and how to promote cooperation in conjunction with productive efficiency using modern technology, particularly in a setting of scarce economic and environmental resources. So, in addition to the interest that the local Jewish settlers in Israel had in the success of the kibbutz and the moshav and that of the Jews of Eastern and Western Europe who had contributed ideas, people, and financial support to the settlement movements, there was also interest on the part of many developing nations and international development agencies subsequent to the establishment of the State of Israel. Israel was fortunate by that time in having at hand a substantial number of skilled observers, social scientists, who could report on the progress of the various types of settlements and who were encouraged to do so (Weitz 1971).

For a time, both the kibbutz and the moshav were intensively studied (see Chapter 2). Many of the studies took the form of very close observation of individual communities (participant observation), entailing residence by the observer in the community and detailed descriptions of the actions and interactions of community members. These ethnographic descriptions provide an extraordinary baseline from which to examine social change. The restudies described in this book were able to take advantage of a wealth of excellent observation in the 1960s and 1970s, in some cases by the same authors.

Restudies of small communities are rare in themselves; a restudy by the same investigator is truly an unusual occurrence. A collection of restudies of the same kind of community in the same region was unknown until this book. Here, we are provided with the opportunity to look at a process of deliberate social change over time, up close, from the perspective of individuals, communities, and a nation.

The issues that have come to the forefront are those that preoccupied the investigators and their informants at the time of the study, most specifically, those concerning individualism and cooperation. While this was an issue for the moshavim, it was also an issue for vast parts of the wider world as Soviet and Eastern European socialism and communism were crumbling. Thus, the outcome of the Israeli settlers' efforts to create institutions of cooperative living based on socialist ideas is of far more than local interest. Briefly, the moshav (and the kibbutz) appeared to be in crisis on a number of fronts—indeed, in the view of some, to be proving failures. Critics argued that the moshav as a form of organization was partially responsible because it led to mismanagement of resources. They also argued that state support of moshavim encouraged such mismanagement (Sherman and Schwartz 1991). On a deeper level, there were doubts about the viability of social organizations dependent on voluntary cooperation. Moshav members (perhaps like all Israelis) appeared to be becoming more individualistic, selfish, and self-centered than they had been in the past. Individualistic tendencies were seen at one and the same time as the source and symptom of the failure of socialist institutions.

The restudies in this book examine the questions of whether and to what degree the moshav members have become more individualistic, why and how and to what effect, and the relationship between changes at the community level and changes at the more inclusive national level. Before examining these issues, we should lay out some basic background for readers unfamiliar with the moshav and its history.

## What Is a Moshav?

Very simply put, a moshav is a village—a small rural residential community—organized as a legal corporation with multiple functions. The corporation (*aguda*) has certain resources, principally land and water, and the ability to take and make loans. It serves as the manager of village functions, as a mutual assistance society, and as a multiservice agricultural cooperative. Membership in a moshav entitles one to a place to live, use of a parcel of land the same size as that of other members, a share in cooperative enterprises, access to the facilities and services of the cooperative, the support of other members in time of need, and a vote in decisions made by the cooperative. It entails an obligation to participate in cultivating the land, to help other members in need, and to cooperate with others in carrying out the decisions of the cooperative, such as to market produce cooperatively rather than individually. For the past forty years, the moshav has been the most common form of rural organization for the Jewish population of Israel. It has been used by the Israeli government and its affiliated agencies as a principal means of absorption of immigrants into the countryside and as an instrument of economic development of rural areas (Willner 1969; Klayman 1970). Thus, the moshav has simultaneously served to meet the needs of both its immediate members and the State of Israel. Sometimes these two tasks have appeared to be in conflict, and each, individually, also contains contradictions.

The moshav, as an institution, though it refers to the individual incorporated village, entails a host of other institutions which have grown in conjunction with it. These institutions are extralocal organizations that serve groups of moshavim in various ways (e.g., marketing cooperatives). Just as the role of the state must be taken into account in any discussion of the moshav and its history, so must the development and role of these institutions, which have received considerable attention in the literature.

## History of the Moshav

The history of the moshav as an institution has been recounted so many times and with such eloquence that it is hardly necessary to do more than provide a brief sketch here. The main principles of the moshav were delineated in 1919 by a pioneer of the second modern wave of immigration to Israel, Eliezer Joffe, and put into effect with the establishment of the first two moshavim, Nahalal and Kfar Yehezkiel, in 1921. In contrast to the kibbutz, which, until the establishment of the State of Israel, was the most common form of Jewish rural settlement, each member of the moshav would hold and operate an individual farm. The work was to be done solely by the holder, so the size of the farm had to be small enough for this but large enough to provide support for the holder's immediate family. The holding would be owned by the Jewish National Fund and leased to the holder on favorable terms. Marketing, economic, and municipal services would be handled cooperatively by and for the membership. Finally, comprehensive arrangements were made to ensure mutual assistance among the members.

Clearly, the moshav was intended as a compromise between the individualistic and cooperativistic ideals of its founders. On the one hand, they did not wish to depend on one another for production, but on the other, they also did not wish to harm one another through competition, nor did they wish to abandon responsibility for one another's welfare (or forsake the support of the group). They adhered to certain other socialist–communist principles as well, including a rejection of private property in land, rejection of the exploitation of laborers, along with acceptance of the ideal of the dignity of labor through the principle of self-labor, members being required to work their own farms.

Aside from some obvious problems in adhering rigidly to any set of ideals in the practical context of farming (such as the complete ban on hired labor—something the moshav movement has never been able to implement), the tensions inherent in the compromise between individualistic and cooperativistic tendencies have persisted from the start. Some members of the communities have leaned toward more individualistic interpretations of the principles, and others, toward more cooperativistic ones. Sometimes this has been a source of conflict, and it always has entailed negotiated compromise over specific issues as they have arisen, particularly as agriculture and the Israeli economy have undergone development and quite dramatic change.

## SOURCES OF DIFFERENTIATION

### Immigration and Ethnicity

The most dramatic change of all followed the establishment of the state, namely, the subsequent influx of huge numbers of immigrants, not only survivors of the Holocaust (whose cultural origins were not so different from the Jewish population of prewar Palestine) but mass immigration from Third World countries, largely Arab ones. For a variety of reasons, some still remaining to be explored, the moshav was selected as the rural institution most appropriate for the settlement of these latter immigrants. With this decision, the moshav changed from being the minority alternative to the kibbutz, to being the most common form of Jewish rural settlement. But the settlement of the new immigrants from North Africa and the Middle East in the moshavim set up new dichotomies, the most important of which are the overlapping distinctions between the "veteran" and the "new" moshavim, and the ethnic distinction between the moshavim whose origins are European (Ashkenazi) and those which are Eastern (Sephardi). Underlying these distinctions are political and economic ones: The new moshavim tended to be not only of Eastern origin, but to be poorer in resources, poorer in economic performance, located in geographically more marginal areas left behind by the earlier settlers, and, until recently, virtually without political influence either in the agricultural sector or at the national level. In the literature on the new moshavim, the ones which received attention in the 1960s are those of Eastern (in other words, Third World) origin, whose ethnicity was considered to be a major issue. It is these communities that were regarded as "reluctant pioneers" and who were subject to the imposition of not only alien ways of life, but alien ideology. Their struggle was depicted as one to learn to adapt to modernization, namely, Israeli bureaucracy, as well as to socialistic ideals of cooperation; subsequent times revealed a different parallel struggle of resistance against the hegemony of Ashkenazi culture and political domination by the Ashkenazi population as well. The tension between expressions of individualistic tendencies and cooperativism in the moshav was made considerably more complex by this development.

### Quotas

Despite the difficulties, the agricultural sector developed rapidly, with a success that should be reexamined in light of subsequent events. In an effort to manage surplus production, the Ministry of Agriculture centralized its agricultural planning, which in 1964 brought about the implementation of a system of allocation of subsidized production quotas for agricultural products marketed in Israel. Veteran moshavim and kibbutzim, which had had a head start and relatively high levels of production of such desirable products as milk, were allocated very favorable quotas compared to the new moshavim. Both veteran and new moshavim had to deal with decisions about internal allocation of quotas: Were they to be shared equally among moshav members, or divided unequally according to interest, desire, and capability? (Baldwin 1972). A similar challenge lay in the allocation of water for irrigation, which, with the development of agriculture and the provision of modern technology, became increasingly important though it was always necessary and scarce (see Chapter 4). The introduction of quotas both for production and water exacerbated the tensions already discussed.

## FURTHER CHANGE

The Six Day War in June 1967 brought a further important change. The conquest of the West Bank and Gaza achieved safer borders and at the same time made accessible an enormous pool of cheap labor—the Arab occupants of these territories including landless refugees. A surge in dependence upon hired labor and the fact that it was Arabs who provided it altered both attitudes toward and the practice of self-labor in the moshavim (see Chapter 10). A change in the borders had a very important impact on many of the moshavim as they shifted from being on the frontier to being incorporated in the urban centers as suburbs (see Chapter 9).

Less dramatic in its impact in that the effects appeared to follow more slowly and subtly was the election in 1977 of the Likud Party to the government of Israel, the first non–Labor Party government since the establishment of the state. While the Labor Party had long been a staunch supporter of the rural institutions—the moshavim and the kibbutzim—the Likud undermined that support and began to direct its attention to settlements in the occupied territories rather than to the needs of the established communities of Israel proper. The consequences were, of course, an erosion in the power base and the economic support of these communities (Sherman 1980). During this period, first runaway inflation, then harsh austerity measures resulted in important economic changes in the moshavim.

The early 1980s saw further major changes, in some cases the bringing of earlier tensions to the breaking point. In 1985 and 1986 came the collapse and bankruptcy of the regional purchasing cooperatives which had served as banks for the moshavim (Schwartz, Marvid, and Arian 1988). This brought to a head the problem of mountainous debt in the agricultural sector. The debt crisis was complex indeed. It was an outcome of the inflationary period which encouraged debt, followed by the austerity measures, which led government to refuse to bail out the rural institutions. It was also a consequence of a global agricultural crisis which entailed costly technological investment, surplus production, and gluts on the market, and resulted in disaster, particularly for small-scale farmers worldwide. For many moshavim in the latter part in the 1980s and early 1990s, the strains led to breakups in the moshav associations for all but municipal functions as members became unable or unwilling to commit themselves to mutual assistance, mutual responsibility for one anothers' debts, or cooperative marketing. Further, when it appeared that the individual small farm was not economically viable or worthwhile, many moshav residents abandoned farming and rented out (or sharecropped) their land to others who expanded the size of their agricultural operations to an economically viable scale.

## SUCCESS OR FAILURE?

Thus, while the moshav remains in the vocabulary of the Israeli rural sector, it has unquestionably changed its reality. In the early 1990s, some were ready to declare the moshav a magnificent failure, while others claimed it had been hugely successful but had simply outgrown its embryonic form. For some, the apparent shift toward individualism (whose meaning requires some exploration) was a grave disappointment, a sign of human and social flaws. Others felt that cooperation and communalism had fallen victim to larger economic and political forces which made the attainment of lofty ideals impossible. Still others felt that this shift was a triumph over a bad idea, socialism. These found support for their convictions in the collapse of socialism in Eastern Europe and the Soviet Union. Finally, there were those who felt that the moshav was a success because it was a good idea for a while, but now the old ideas and ways of being had to give way to new ideas and methods. It is still too soon to say what the new ideas and methods are, but it should be clear from this brief history that many of the issues that confronted the moshavim in the 1990s had roots much earlier on, especially the issues related to cooperation and individualism and equality and difference. The particular expressions of these issues were always grounded in the immediate context of their times, but in some ways, these expressions were also cumulative in that there was never a final resolution but just new additions. Perhaps this is the way that society by design develops into natural society.

## REFERENCES


Baldwin, E. 1972. *Differentiation and Co-operation in an Israeli Veteran Moshav.* Manchester: Manchester University Press.

Klayman, M. I. 1970. *The Moshav in Israel: A Case Study of Institution-Building for Agricultural Development.* New York: Praeger.

Schwartz, M., A. Marvid, and O. Arian. 1988. *The Organizations: Three Moshav Purchasing Organizations before and after the Crisis.* Rehovot: Settlement Studies Center.

Sherman, N. 1980. "From Government to Opposition: The Rural Settlement Movements of the Israel Labor Party in the Wake of the Election of 1977." *International Journal of Middle East Studies* 14: 53–69.

Sherman, N., and M. Schwartz. 1991. "The Effect of Public Financial Assistance on the Management of Moshav Economic Affairs." *Human Organization* 50(2): 163–172.

Weitz, R. 1971. "The Role of the Sociologist in Israel's Rural Development." In *Rural Settlements of New Immigrants in Israel,* O. Shapiro, ed., pp. 15–28. Rehovot: Settlement Studies Center.

Willner, D. 1969. *Nation-Building and Community in Israel.* Princeton: Princeton University Press.

PART I

# ORGANIZATIONAL ENVIRONMENT: BEGINNING AND LATER DEVELOPMENT

Chapter 1

# Moshav and Kibbutz in the Anthropology and Sociology of Israel

*Moshe Schwartz*

*Israel's collective and cooperative settlements (the kibbutz and the moshav) were among the major objects of study of Israeli sociology and anthropology until the mid-1970s. This chapter explains how this situation came about, how it was maintained for more than twenty years, and why after those twenty years the two settlement types lost a great deal of their attraction for social researchers. Interest in moshavim and kibbutzim was rekindled by the financial and organizational crisis in the second half of the 1980s and by the rapid changes which followed in its wake.*

## THE KIBBUTZ AND THE MOSHAV AS MAIN TOPICS OF ISRAEL'S SOCIOLOGY AND ANTHROPOLOGY

### Scholars and Publications

Until the mid-1970s, Israel's two main forms of planned settlements, the kibbutz (collective) and the moshav (cooperative), were the main objects of anthropological research in Israel. More than 50 percent of the scholars and more than half of the publications listed in Handelman and Deshen's bibliographical essay, *The Social Anthropology of Israel* (1975), dealt with the two settlement types (see Table 1.1), as did the first two chapters of the essay (out of five).

About 50 percent of the sociologists who published articles or books about the Israeli society devoted some of them to the kibbutz and moshav, and about half of their publications dealt with one or the other. This is shown in Table 1.2. It is based on Weller's *Sociology in Israel*, which also devoted a separate chapter (out of eight) to the kibbutz, though none to the moshav, which is less studied by sociologists than by anthropologists. However, the first chapter of

**Table 1.1**
**Kibbutz and Moshav in the Anthropology of Israel, 1948–1974**

| | Kibbutz and Moshav | Total | Kibbutz & moshav as % of all topics |
|---|---|---|---|
| Scholars from all social sciences | 27 | 47 | 57% |
| Published by the above | 77 | 154 | 50% |
| Anthropologists | 17 | 31 | 55% |
| Publ. by anthropologists | 57 | 117 | 49% |

Calculated on the basis of Handelman and Deshen (1975).

**Table 1.2**
**Moshav in the Sociology of Israel, 1948–1973**

| | Kibbutz and Moshav | All topics | Kibbutz & moshav as % of all topics |
|---|---|---|---|
| Scholars from all social sciences | 19 | 35 | 54% |
| Published by the above | 52 | 124 | 42% |
| Sociologists | 14 | 29 | 48% |
| Publ. by Sociologists | 36 | 106 | 34% |

Calculated on the basis of Weller (1974).

the book, "The Absorption of Immigrants," includes a section dealing with the moshav.

### The Kibbutz and the Moshav in Israeli Academia

The study of the kibbutz and the moshav enjoyed academic respectability. In 1972, three out of the four Departments of Sociology and Anthropology offered courses on the subjects of the Sociology of the Kibbutz, Sociology of the Moshav, or Rural Sociology. The exception was the Department of Sociology and Anthropology at Bar Ilan University, a religiously oriented institution not far from Tel Aviv. At Bar Ilan, two economists, Yehuda Don and Shlomo Eckstein, specialized in the two settlement types.

The Department of Sociology of the Hebrew University in Jerusalem was established much earlier than the others, just as the Hebrew University was. It was also the first academic institution to show interest in these two settlement types. The sociologist S. N. Eisenstadt, who headed the Department for more than thirty years, claims that he and Yonina Talmon-Garber, another prominent sociologist, were already studying the moshav as early as the mid-1940s, long before they completed their Ph.D.s. However, due to World War II, that research was never published (see introduction to Talmon-Garber 1972). Later on, in the 1950s and 1960s, Eisenstadt published several books and articles about the absorption of immigrants in the new moshavim, while Talmon-Garber made the kibbutzim her main topic of research. Joseph Ben-David, a third senior member of the Department, published a historical analysis of the growth of the two settlement types and of their historical role in the development of the country. He also dealt with the role that kibbutz members played in politics as members of the political elite (Ben-David 1964).

In addition, the Department of Sociology at the Hebrew University held a permanent research seminar dealing with The Sociology of the Kibbutz. Of thirteen full-time department members in 1972, three mentioned the study of the moshav or the kibbutz as one of their special interests (Cohen, Goldberg, and Weintraub). Weintraub was undoubtedly the most prolific writer about the moshav. Indeed, out of 114 sociological publications in English or French on the subjects that are mentioned in Meshorer's bibliography (1987), twenty-eight (i.e., almost 25 percent) were written by Weintraub or in collaboration with him. The period during which he published about the moshav is unusually long. It lasted for thirty years, from 1952 until 1982, that is, shortly before his mortal illness. There were three other members of the Department of Sociology of the Hebrew University who also had published papers about the two settlement types.

There even was an official linkage between the Department of Sociology of the Hebrew University and the Land Settlement Department of the Jewish Agency. The Settlement Department had established most of the moshavim and kibbutzim of the country and was still in charge of setting up new ones and taking care of them until they could take care of themselves. The linkage between the Settlement Department of the Jewish Agency and the Department of Sociology of the Hebrew University was the Council for Social Issues, which was set up in 1959 and headed by the chairmen of both institutions, Eisenstadt and Weitz (Shapiro 1971; Weitz 1962). It should not be assumed, however, that Eisenstadt was just as interested in the study of the moshav as Weitz. Judging by his list of publications in the early 1960s, Eisenstadt had more interest in Israeli society in general, or in historical and political sociology, than he had in the moshavim. Indeed, he may already have lost interest in the latter subject (Weingrod 1994).

In 1972, two of the fourteen full-time members of the staff at Tel Aviv University, Moshe Shokeid and Israel Sheffer, listed respectively the moshav or the kibbutz as their main topic of interest, and three other members, Shlomo Deshen, Shimon Peres, and Ephraim Yuchtman-Yaar, had also published articles about the two settlement types. Although some members of the Tel Aviv department had studied at the Hebrew University under Eisenstadt, they were no longer sure that they agreed with his or any other functionalist theories. Even their approach to kibbutz and moshav studies was definitely different from that of the Hebrew University group. The anthropologists in the Tel Aviv department had been influenced by their doctoral advisor, Max Gluckman, and by some of his outstanding students, such as Victor Turner. Gluckman and Eisenstadt fought several times publicly about the latter's concept of modernization and its use in the study of the moshav. Gluckman made no secret of being furious about the matter (Nevo 1993b).

That same year (1972) there were four kibbutz, moshav, or cooperative specialists among the thirteen staff members in the Department of Sociology and Anthropology of the University of Haifa. Joseph Shepher headed the Research Institute of the Ihud Hakvutsot Vehakibbutzim, one of the three major federations of kibbutzim linked with the Israeli Labor Party. Menahem Rosner, the head of another federation's research institute, was to join the staff of the Department of Sociology and Anthropology of the University of Haifa at the end of the 1970s.

The Ben-Gurion University of the Negev in Beer Sheva tended to specialize in engineering and in the physical sciences more than in social sciences. Nevertheless, in 1972 this university already had a Department of Behavioral Sciences (where psychologists made up the largest group). Some of the sociologists and anthropologists at the same university published research on the kibbutz (Antonovsky and Arian 1972) and the moshav (Weingrod 1966).

The Technion—the Technological Institute of Israel—was more exclusive in its orientation to engineering and architecture than Ben-Gurion University. Indeed, the Technion did not have a separate Department of Sociology or Behavioral Sciences. The sociologists employed in 1972 by that university specialized in the sociology of work and organizations, or in Urban Studies, and neither was particularly interested in the kibbutz and the moshav.

On the whole, the kibbutz and the moshav were important topics of study in the Social Science Departments of Israeli universities in 1972, though not as important as indicated by their place in anthropological and sociological publications until the mid-1970s. Specialists of the kibbutz and the moshav constituted, on the average, less than 15 percent of the staff of Israeli universities. Besides, most academic specialists in the two settlement types had additional objects of interest. The fact that there was less interest in the kibbutz and the moshav in academia indicated that they were about to lose some of their popularity as objects of research.

## Interest in the Study of the Kibbutz and the Moshav among Their Members

On the whole, the sociological and anthropological study of the kibbutz and the moshav awakened interest among the members of both settlement forms, though more in kibbutzim than in moshavim. Baldwin (1972) and Nevo (1978) report mild curiosity about their research in the veteran moshavim they studied, whereas members of kibbutzim had from the very beginning strong feelings about sociology: They were often irritated by Landshut's study (1944), objected to Spiro's (1951) and Rosenfeld's (1956) findings, and held passionate arguments between opponents and supporters of Talmon-Garber's methods and findings.

Indeed, a substantial number of kibbutz members followed in the footsteps of Talmon-Garber's students and became professional sociologists or anthropologists, often specializing in the study of their own society, the kibbutz. There was no parallel phenomenon in the moshav: Few veteran moshav members or sons of veteran moshavim made sociology or anthropology their professions, Shapiro probably being the most notable exception.

A possible explanation of the interest of kibbutz members in the study of society is that having failed to change it, they wanted at least to understand it. This could be true not only for the kibbutz joiners but also for the kibbutz-born who inherited from their parents interest in social issues and the urge to understand what had gone wrong in their own society.

## Reservations and Scholarly Arguments around the Study of the Kibbutz and the Moshav

The respectability of the study of the kibbutz and the moshav was not uncontested. Thus, Ben-David (1972) expressed some regret that Yonina Talmon-Garber had spent all of her brief career studying the kibbutz, implying that her great professional gifts could have been better employed.

Besides, some members of the social science community became disaffected with some of the research done at the institutes of the kibbutz federations. To those members, that research appeared to spend too much effort to protect the kibbutz against criticism. Thus, it denied inequality between the sexes in the kibbutz (Rosner 1972), intergenerational tensions (Rosner et al. 1990 [1978]), and any serious difficulties in the transition of the kibbutz from agriculture to industry occupationally and economically (Leviathan and Rosner 1980). They also praised unreservedly kibbutz efficiency (Shepher 1970) and democracy (Rosner and Cohen 1980), almost appearing to affirm that in the kibbutz "everything was for the best in the best of all possible worlds." And not all sociological commentators were as gentle as Yuchtman-Yaar in his praise of Leviathan and Rosner's book: "The contributors have a commitment to the kibbutz movement and their data interpretations are often marked by it. But notwithstanding some unwarranted generalizations this collection is a 'must' for students of the kibbutz" (1983).

Among specialists of the moshav, there were disagreements between the students of Eisenstadt who tended to interpret what had happened in the moshavim in terms of theories of modernization and persisted in their interpretation as late as the mid-1960s (Weintraub and Bernstein 1966) and the anthropologists (British—social; or American—cultural) who had qualms about the adequacy of these theories to interpret what was going on in the moshavim and believed that the members of the new moshavim, whatever their ethnic background, were as capable of economic and political rationality as any other Israelis.

## THE DECLINE OF SOCIOLOGICAL INTEREST IN THE MOSHAV AND THE KIBBUTZ AFTER 1975

Interest in the moshav has been declining since the mid-1970s. A statistical analysis of Meshorer's exhaustive bibliography of the moshav (1987) shows that most of the publications in English or French included in it were published in 1974 or earlier and that very few researchers joined the field after that date. Indeed, it is well known that for any academic specialty the lack of new and committed researchers is an indication of decline. Table 1.3 documents the quantitative decline of the research about the moshav.

A significant proportion of the remaining moshav scholars were those who had been writing about it for twenty to thirty years, sometimes motivated by attachment to their community where they did their doctoral fieldwork, as is the case for some contributors to this book.

The only social research institution where a sizable part of the research was still devoted to the moshav was the Settlement Study Center at Rehovot. Two notable pieces of research about the moshav were published at the Center in the second half of the 1970s: Applebaum and Margolis's book *The Moshav and the Test of Time—Patterns of Change in an Organization* (1979) and Don, Ilan, Bar-El, Bar Hai, Greenspan, and Schnitt's book *Moshav Industrialization* (1977). One of the characteristics of the research done at the center was its interdisciplinary character. In the first publication, one of the authors is a geographer and the other a sociologist, while in the second the senior author is an economist with sociologists on his team. This tendency to do interdisciplinary work in the Settlement Department was to persist in the 1980s.

The one Hebrew University sociologist who maintained contact with the field was the late Dov Weintraub, who did so through his

**Table 1.3**
**Publications in English and French about the Moshav before and after 1974**

| | Publications | % | New Researchers | % |
|---|---|---|---|---|
| 1948-1974 | 101 | 88% | 30 | 89% |
| 1975-1986 | 13 | 12% | 4 | 11% |
| Total | 114 | 100% | 34 | 100% |

Based on Meshorer (1987).

teaching at the Faculty of Agriculture and even more through his involvement as an academic research advisor to the researchers at the Settlement Studies Center.

By and large, the anthropologists rarely returned to the scene of the moshav except for occasional visits, with the exception of Nevo who, being employed by the Settlement Department, continued to devote much of her attention to it. Similarly among the sociologists of the Settlement Department, the proportion of anthropologists dwindled. What was expected from sociologists was not so much to make brief anthropological surveys of settlements or settings, but rather to evaluate candidates for membership in the settlements. More generally, their role was increasingly bureaucratic and their involvement in research tended to decrease with the passage of time.

### Sociologists of the Kibbutz

Sociologists of the kibbutz, most of them kibbutz members, have remained a sizable group. Indeed, their numbers were on the increase, as well as their productivity. The Israeli Sociological Association has even devoted the second yearly thematic collection of articles in its series *Studies of Israeli Society* to the Sociology of the Kibbutz (Krausz 1983). However, most of the articles were originally published before 1974, contrary to all other themes of Israeli society. Table 1.4 is given for comparative purposes.

**Table 1.4**
**Dates of Publication of Articles Reprinted in the *Studies of Contemporary Israeli Society* Series**

| | 1974 and before | | 1975 and after | |
|---|---|---|---|---|
| Migration, Ethnicity and Community (1980) | 4 | (40%) | 6 | (60%) |
| Kibbutz (1983) | 12 | (57%) | 9 | (43%) |
| Political Sociology (1985) | 1 | (6%) | 15 | (94%) |
| Education (1989) | 0 | | 20 | (100%) |
| Health and Health Care (1991) | 0 | | 25 | (100%) |
| Women (1993) | 0 | | 26 | (100%) |

Dates in parentheses refer to the publication of the collections.

Furthermore, most of the sociological research about the kibbutz went over to the research institutes of the kibbutz federations, and only a few non-kibbutz members such as Ben-Rafael and Kressel were still involved in it. Courses about the sociology of the kibbutz became more scarce at Israel's universities even though there was some revival of interest in the kibbutz as a result of the crisis.

## WHY WAS THERE SO MUCH INTEREST, AND WHY HAS IT RECEDED?

### The Role of Moshavim and Kibbutzim in Nation-Building

During Israel's first decade of statehood, the moshavim, new and veteran, filled essential functions in nation-building (Willner 1969). The new moshavim absorbed large numbers of new immigrants and attempted to subject them to a far-reaching social transformation, making them into farmers and pioneers (Weingrod 1966); they also dispersed the population and contributed to the security of the country's borders. The veteran moshavim tremendously increased agricultural production, providing much of the food that the country desperately needed, especially during the beginning of the decade. The new moshavim also made a tremendous contribution to the country's agricultural production (Schwartz and Giladi 1993).

While the kibbutzim failed to absorb new immigrants to any significant extent, they did play an important role in the country's food production, as well as in the dispersal of its population (Schwartz and Giladi 1993) and the building of its industry (Barkai 1977). They also played an important role in the country's defense, probably more than the moshavim, in two ways: One is that they were often set at what were at the time the most dangerous borders of the country—the Gaza Strip and the Jordan Valley where the kibbutzim were exposed to Syrian bombardments. Another way was their disproportionate contribution to staffing Israel's elite military units and to filling the ranks of the young officers. Thus, it is no accident that a settlement form representing less than 25 percent of the country's population provided it with its two last chiefs of staff, both of whom were born in kibbutzim. Nevertheless, the contribution of the kibbutzim to the security of the country in statehood years was much less than it had been in the late 1930s and 1940s when they had borne the brunt of the national security of the Jewish Settlement in Palestine.

All in all, the kibbutzim and the moshavim were at the center of the country's national revival, and they had problems to cope with. Anthropologists and sociologists were drawn to bear witness to the national epos and to contribute their share to the national revival of the Jewish people in its newly established state by helping to solve some of its problems.

### The Ideals of the Kibbutz and the Moshav

The ideals of the kibbutz and the moshav were in tune with the socialist leanings of the post-World War II generation of Jewish youth, including the social scientists. Thus, they were tempted to study Israeli society and "the greatest social experiment of all times," with its complement of human drama and "venture in utopia." For an additional interpretation of what motivated some of the American social scientists, see Weller (1974), based on Rosenfeld (1958).

### International Sympathy for Israel

Until the Sinai Campaign and even until the Six Day War, the young state of Israel was an object of sympathy and curiosity in the Western world, especially among intellectuals. The memories of the Holocaust, the victories of the Independence War pictures as the victory of David over Goliath, the impressive agricultural successes of Israel in a world still hungry, and the attempts to "make the desert flourish" elicited a mixture of benevolence, admiration, and curiosity. The kibbutz and the moshav, even more than other subjects connected with the young State of Israel, elicited sympathy and curiosity in the international scholarly community. Thus, students specializing in the two kinds of settlements had some reasonable expectations that their books or papers would be published.

These positive attitudes toward the State of Israel lost much of their potency as it started to project an image of strength after the Six Day War, losing much of the popularity that intellectuals of the West tend to reserve for the underdog (see Table 1.5). Thus, the expectations of publishing things about Israel diminished somewhat, unless they were things, such as the study of the Black Panthers, that had a critical view of Israeli society, or at least stressed its problems as much as its successes (Cohen 1982). Also, agriculture and the countryside to which the two settlement types belonged lost much of their urgency in a world that had solved some of the problems of food production.

**Table 1.5**
**Hebrew Compared to English or French Publications about the Moshav**

| | Hebrew | Period as % of Total |
|---|---|---|
| Until 1955 | 5 | 2.5% |
| 1956-1960 | 24 | 12.0% |
| 1961-1965 | 45 | 22.5% |
| 1966-1970 | 34 | 17.0% |
| 1971-1975 | 35 | 17.2% |
| 1976-1980 | 28 | 14.0% |
| 1981-1986 | 33 | 16.0% |
| Total | 204 | 100.0% |

## Research Methods

Anthropological research methods had already been used as early as the 1930s to study small communities in modern societies and had turned out to be well suited for the job, producing classical community studies such as Arensberg's *Irish Countryman* (1937), as well as methodological pieces such as his article "The Community Study Method" (1954), often quoted in introductory and research methods textbooks. Other well-known community study classics such as the first volume of Warner and Lunt's *The Social Life of a Modern Community* (1941) Yankee City Series and Whyte's *Street Corner Society* (1943) were much influenced by Arensberg.

Later on, Elisabeth Bott (1957) discovered that the methods of anthropology could work in the big city if you focused on families and their social networks. Talmon-Garber applied insights from Bott's theory of networks to the study of the kibbutz, while Shokeid applied them to the study of the moshav. It makes no difference for the present study that the findings of Bott have been heavily contested by researchers and that they are not accepted anymore by most. It is still the case that they played an important role in the history of sociology and anthropology in general and in that of Israeli sociology and anthropology in particular.

## The Anthropologists as Would-Be "Strangers"

While the methods of anthropology had been used for the study of communities in modern society, Ellen (1984) has rightfully pointed out that the preference of anthropologists has remained for the exotic, the unusual: a youth group in an Italian ethnic neighborhood (Whyte 1943), group of lower-class Blacks (Liebow 1975), and a Central American small town (Redfield 1959; Nash 1962). From that point of view, the study of the new moshavim and especially that of the new moshavim populated by Oriental Jews (especially Jews from Moslem countries) allowed Israeli anthropologists to "have it both ways"—at one and the same time they could study things central to their own society and people foreign to them, thus joining the strange and the familiar and being able to say that the strangers are not so different after all, that they are fellow human beings and fellow Jews.

Sociologists (such as Eisenstadt, Talmon-Garber, and Weintraub) added questionnaires to fieldwork. This turned out to be especially useful in the study of the kibbutz: Some of the researchers were or had been kibbutz members and knew the right questions.

## Funding and Career Opportunities

Jobs and research grants were available for those willing to study the moshav thanks to the funding provided by organizations sponsored by Jewish philanthropists who took those two settlement types seriously as contributors to nation-building and, to some extent, also as social experiments. Thus, researchers could make a living and build a professional career for themselves while studying the moshav. Indeed, few other sectors, if any, provided the opportunity for promising university graduates to do research work in the social sciences and for professional scholars to get research money from a cooperative and open-minded establishment. The Land Settlement Department played an essential role in the process.

## The Role of the Land Settlement Department

The Land Settlement Department of the Jewish Agency (later on, the Settlement Department) played an important role not only by providing jobs to researchers (and also funds through the Settlement Study Center and through the faculty of agriculture which also received research funds) but also by offering them freedom of research and publication. The Land Settlement Department, established in the early 1920s, had established the kibbutzim and most moshavim and provided them with housing and initial productive investments (while the Jewish National Fund provided the land). The founder of the Department had been sociologist Arthur Ruppin, who believed that the social sciences might be able to assist in its settlement effort. Indeed, in 1939 he helped initiate and fund the first sociological study of the kibbutz by Dr. Siegfried Landshut. That study was completed in 1941 and published in Hebrew in 1944 by the Institute for Zionist Education for the enlightment of the general public. This pioneering study had some influence on later students of the kibbutz (Talmon-Garber quoted it), but given the circumstances of its publication during World War II, less influence than it might have had otherwise. Landshut's ideas were later on to become crucial in the sociology of the kibbutz, but his methods were not as systematic as those of his successors. Besides, he attempted to preach at the kibbutz members what he as a socialist felt that they should do, and this attempt made them feel that he was superficial and presumptuous (Talmon-Garber 1972).

Raanan Weitz, Director of the Settlement Department in the 1950s and early 1960s (under the chairmanship of Levy Eshkol) and later Chairman of the Department, introduced sociologists into the Department. He said that when he started to set up hundreds of new moshavim he believed that anthropologists might be able to help him and he invited Margaret Mead to advise him. She refused, he said, claiming that it was his job to make the mistakes and hers to record them for posterity.

Weitz persisted in his attempts to harness sociology and anthropology to the settlement effort and ended up by hiring Dorothy Willner, who had studied under Robert Redfield at the Anthropology Department of the University of Chicago. She became the first Settlement Department sociologist in 1955, the year when her teacher's book, *The Little Community*, was published. In 1959, Alex Weingrod received his Ph.D. in Anthropology at the University of Chicago after doing his field work in the moshav Tidhar in the Negev and became the first head sociologist of the Department. In his new function, he set up "district sociologists" (Shapiro 1971), among them, Ovadia Shapiro in the Jerusalem Region, Naomi Nevo in the Negev, Rivka Rahat in the North, and Zipora Stup in the Lachish Region, and also Moshe Minkovitz and Shlomo Deshen. Later on, when Shapiro became head sociologist, Nevo was replaced by Moshe Shokeid and Stup was replaced by Shlomo Deshen. Shokeid and Deshen went on, with the encouragement of Raanan Weitz, to work on Ph.D.s in Great Britain at the Department of Anthropology of Manchester University. Another American anthropology student who did his field work on an Israeli moshav was Harvey Goldberg, who completed his Ph.D. at Harvard at the Department of Social Relations and later published his thesis in book form (Goldberg 1972). It should be noted that most of the Settlement Department sociologists, at least in the first generation, called themselves anthropologists and worked for doctorates in anthropology though the department insisted on calling them sociologists (Weingrod 1993).

### Sociologists in the Settlement Process

The need for sociologists was not felt during the years of mass settlement (the first five years of the State of Israel) but were considered necessary later on, especially between 1955 and 1965, when it was hoped that they would help to solve the problems which had emerged, such as the factional struggles that plagued the immigrant moshavim and that were found very disturbing by the officials of the Department. One point on which the view of sociologists was accepted was that ethnically homogeneous moshavim might be preferable to ethnically diverse ones: Sociologists believed that the drastic change the new immigrants had to undergo would be cushioned if they lived among people they knew or at least among people who spoke the same language. Some sociologists also believed that homogeneity might soften factional strife and lessen the likelihood of stalemate, thus enhancing the chances of development. About the latter view there was indeed considerable controversy, and it is by no means clear that it was always correct. Finally, the Settlement Department officials believed that the anthropologists might help the planners to take into account properly the cultural backgrounds of the new immigrants so as to avoid costly mistakes.

The Department considered its sociologists as minor officials and expected loyalty. All in all, the sociologists accepted that role although it did create some problems and dilemmas between the role of the professional and that of the bureaucrat, as well as between loyalty to the officials above them and to the communities they studied (Nevo 1993a). The dilemmas were mitigated by the support that the Department sociologists received from Weitz, by professional respect they enjoyed among most of the officialdom, and also by the developed spirit of cooperation and professional support which evolved between the members of the team and which continued to characterize it for many years.

It should be noted that a major reason for the introduction of the sociologists was to help solve the problems of the new moshavim, many of which turned out to be incapable of proper functioning as small holder cooperatives on the model of the veteran moshavim. In particular, these new moshavim could not properly assist their membership in current functioning or in agricultural investment. Hiring the sociologists to try and put things right was only one of the solutions tried. Another one was the development of regional economic organizations, which often took over some of the most important functions of the newly established moshavim, such as assisting current agricultural functioning by providing inputs and processing produce and also assisting the development process by helping to muster credit for investment purposes (Schwartz, Marvid, and Arian 1988).

The Land Settlement Department of the Jewish Agency also eased the entrance of social researchers to the new moshavim who were still under its aegis. These new moshavim were not in a position to refuse entrance to a researcher employed by the Department or recommended by it. In the case of the veteran moshavim, their management committees played the role of gatekeepers, and, for the kibbutzim, that role was played by both the Settlement Federations and the local leadership of every kibbutz. However, it is the Settlement Department that set a model of openness to social research even though such research sometimes criticized its own policies and deeds.

*The Attitudes of the Researchers: Involvement, Problems, Disappointments, Ambivalence, and Bearing Witness.* None of these researchers showed indifference to the goals that moshavim and kibbutzim had been set up to accomplish, nor to the hardships undergone by their members. Sometimes the researchers expressed their sympathy or admiration for the Zionist enterprise (as in Gluckman's introductions to Shokeid and Baldwin's books), while at other times they found other ways to convey their solidarity, such as Kushner's dedication of his book to his "cousin" Moshe Shertok (Kushner 1973). Sometimes the attentive reader may perceive the anthropologist's effort to avoid saying anything that might tarnish the image of "his community" (Baldwin 1972).

Even where the researchers express ambivalence (Spiro 1956) or disappointment (Rosenfeld 1951), these feelings only demonstrate their identification with those they studied and with the goals they pursued. The disappointments due to the inability of communities to live up to their ideals were of course more frequent, or at least more serious, regarding the kibbutz rather than the moshav. After all, the social goals of the moshav were not quite as ambitious as those of the kibbutz. Furthermore, the above statement only holds true for the studies of the veteran moshavim: As far as the new moshavim were concerned, anthropologists were well aware that no ideological commitment was to be expected among their members, who after all only were "reluctant pioneers" (Weingrod 1966).

Still, in the veteran moshavim and in the kibbutzim, there were inconsistencies between words and deeds in the matter of mutual assistance and cooperation as increasing branch differentiation (Baldwin 1972; Shoresh 1988; Zussman 1988) and disparities of wealth made reciprocity increasingly difficult to maintain.

Besides, there were family dramas connected with the handing down of the farm to one of the sons while the parents were still alive, often of working age, and while the other sons and daughters had to leave the moshav and sometimes forego an inheritance (the law required that the farm be handed down to one and only one of the sons or daughters of the small holder). And there was factional strife with the participation of extended family groupings, described by Shokeid (1985 [1971]) for the new moshavim and by Nevo (1978) for the veteran, a strife in which the "winner takes all" strategy was often adopted by the faction that won the elections (Schwartz, Finkel, Bar-El, and Nesher 1986). All of these provided plenty of stories for drama, and the books about the immigrant moshavim have no lack of it.

The story of the new moshavim reached epic dimensions: mass settlement, the ingathering of the exiles, great social and economic transformations, personal hardship, personal and ethnic humiliation for those who often had to adopt ways foreign and threatening to them against their will, and the painful experience of generations of immigrants watching the gradual estrangement of their own offspring.

Both the kibbutz and the moshav could be considered success stories. Martin Buber, the founder of the Department of Sociology at the Hebrew University, called the kibbutz "an experiment that did not fail" (Buber 1949), and Gluckman, less shy of superlatives, wrote that "There has probably been no greater pioneering social enterprise in the history of mankind. . . . Yet save in very difficult ecological pockets [Gluckman may have had in mind the mountain moshavim] the experiment has proven strikingly successful" (Shokeid 1985).

*The Contribution of Research: Quality, Theoretical Orientations, Changes over Time, and Solutions to Practical Problems.* The combination of an adequate intellectual tradition, sensitizing the researchers to the proper issues, suitable funding, research techniques appropriate to the matter under investigation, and deep involvement of scholars who felt that they were witnessing history in the making yielded some very interesting papers and, a few years later, superior monographs about moshavim.

In the 1950s, there were already some interesting papers about the moshavim of the Negev, such as Weingrod's papers about Tidhar, Nevo's paper about Brosh and Shikma, and Shapiro's Kiryat Yearim—interesting because, on the whole, the most interesting papers were about the moshavim of the Negev. The sociologists and anthropologists of the Jerusalem region (Jewish Agency) did not produce as much perhaps because the Hebrew University Department of Sociology was too close for comfort.

It takes some years from the moment a scholar is plunged into an entirely new field until he or she becomes capable of writing papers about it, and then until he or she reaches peak competence and learns how to distinguish that which is essential from that which is not. This statement is even truer when the field is not only new for the scholar but also for the community of scholars of which he or she is a part. Thus, there was a time lag between the beginnings of the anthropologists' involvement with the moshav and the time their publications started to come out. Those who later on wrote doctoral dissertations and monographs about the moshav started to publish reports and scholarly papers in the late 1950s and early 1960s. Some more time had to elapse until they wrote their dissertations and made them into publishable monographs, some of which may be considered as their best achievements.

While the first anthropologists joined the Settlement Department in the second part of the 1950s, it is in the second part of the 1960s and the first part of the 1970s that their most important and best monographs about the moshav came out.

It is worth mentioning that a ten-year interval between a research effort and its publication is not unusual in sociological and anthropological literature. The research for the *Yankee City Series*, for example, was performed from 1930 to 1934, while publication started in 1941 and went on until 1947.

The earliest papers and books about the kibbutz still being read are those written in the early 1950s by young but well qualified Jewish American scholars. Landshut's book (1944) is neither available to most nor known and appreciated. Thus, it may be said that the intensive study of the kibbutz started ten to fifteen years or more before that of the moshav. Rosenfeld came first, and Spiro did his field work in 1951 and published his first piece in 1955. Richard Schwartz published his first paper in 1954 and his second in 1955. Diamond came somewhat later and published his first papers in 1957 and his next one in 1975. The socialist or liberal leanings of these researchers, while unpractical for young Americans during the Cold War and positively dangerous in the MacCarthy era, were not unusual among young Jewish intellectuals. They had come to find out how "the other society" worked. Rosenfeld wanted to find out whether stratification was universal, and Richard Schwartz acutely rephrased that problem in terms of the contradiction between the demand for equality and the requirements of efficiency. This problem also preoccupied Yuchtman-Yaar (1972) and, later on, the economist Helman (1985, 1987) and was to become especially important in the late 1980s when economic viability became a prerequisite of kibbutz success. Spiro (1954) wanted to find out whether the family was universal and what happened when you tried to transfer the function of socialization to the community. Diamond (1957a, 1957b) showed that the socialist society of the kibbutz had been created in opposition to the values of the East European shtetel (Zborowski and Herzog 1952), without being fully aware of the similarities it still had with it. And R. Schwartz dealt with issues of legal control in the kvutza (1954) and functional alternatives to inequality (1955).

One of Diamond's research assistants said that Diamond, who had carefully recorded detailed observations about kinship in his kibbutz, could not bring himself to subject them to kinship analysis, claiming that "treating one of the most important experiments of our time as if it were some tribal society" would be tantamount to "desecration" (Stone 1972).

It is also worth mentioning that in the introduction to the first edition of his *Venture in Utopia*, Spiro wrote that if the experiment in the Soviet Union failed, this would have an impact on the kibbutz. He then changed his mind in the introduction to the second edition, stating that he now believed in the resilience of the kibbutz. The readers will have to decide for themselves whether Spiro was right in the first introduction or in the second.

The moshav researchers were altogether fewer than their kibbutz counterparts, and, besides, their institutional affiliations were different from those of the latter. Some were employed by the Settlement Department as district sociologists before starting their graduate studies or soon after it. A few had trained in the United States, such as Willner, Weingrod, and Goldberg, and others, including Shokeid and Deshen, studied at the Hebrew University, received some research training with Talmon-Garber, went on to work for the Settlement Department, and then worked on their Ph.D.s under the supervision of Gluckman at the University of Manchester, where Talmon-Garber had spent some time earlier. They received help from the Bernstein Fund, established to promote research about Israeli society. In the second half of the 1960s, a score or so of doctoral students at the University of Manchester were working on theses about Israeli society under the guidance of Gluckman and receiving help from the Bernstein Fund. Most of them specialized in the moshav.

The anthropological study of the moshav expressed a humanistic yet realistic approach to the understanding of new immigrants and to the analysis of the moshav as a social setting. It was a humanistic approach because it distanced itself from the arrogance sometimes displayed by officials. While truly concerned with the economic welfare of the settlers, those officials often stereotyped them, labeling them "human material" and treating their adaptation problems as if the only important thing was how fast they would be amenable to "modernization" or to the values of the moshav (Schwartz 1995).

The anthropologists resisted the stereotyping of the oriental settlers and demonstrated that they were as rational as anyone else in their behaviors and decision criteria and that the organization of the moshav was not always appropriate for their cultural background.

The Settlement Department sociologists also promoted realistic views about the conflicts between material, symbolic, and political interests in the moshav and moderated the ideological enthusiasms of those who promoted the ingathering of exiles and wanted to put all immigrants together irrespective of cultural background and social structure. Their influence was acknowledged, though it may have been less than they would have liked it to be (Weitz 1962). Still, they have legitimized the natural curiosity of some administrators about the politics of the moshav and the intuitive feeling of those administrators that a better understanding of these politics might help them to better assist the development of the moshavim. However, "a little knowledge is a dangerous thing," and official attempts to intervene in the internal affairs of the moshavim seldom improved the situation (Sherman, Schwartz, and Sternberg 1984). It also appears that the anthropologists helped eliminate the expression of raw prejudice from legitimate public discourse.

Anthropological research about the kibbutz sometimes had a tendency for "debunking" and served as a corrective against its idealization. Kressel (1974, 1983) used detailed ethnographic recording to analyze undercurrents of stratification in the kibbutz in both the production and the consumption sphere and in the new context of rapid industrialization and higher standards of living characteristic of the 1960s. Shepher (1983) showed that the official mechanisms of division of labor (*sidur avoda*) had come to have authority only over the powerless, such as the temporary residents of the kibbutz and some of the women (those who do not have "professional" status). Tiger and Shepher published a provocative book (1975) about the division of labor between the sexes in the kibbutz, claiming that the women cooperated in their own and their sisters' relegation to "feminine" roles.

It is difficult to do justice to the richness of the sociological literature about the kibbutz in a few lines of a chapter dealing mostly with motive. One should mention, however, that some of the best pieces of that literature from the 1960s continued to deal with general sociological issues as the earlier writings had. One should also mention the issues of structural differentiation (Cohen 1976), industralization (Don, Ilan, Bar-El, Bar Hai, Greenspan, and Schnitt 1977), generations (Rosner, Ben-David, Avnat, Cohen, and Leviathan 1990), stratification and reward distribution (Ben-Rafael 1980; Yuchtman-Yaar 1972), and the position of the kibbutz in Israeli society.

However impressive as the contribution of moshav and kibbutz studies until the mid-1970s may have been from both the scholarly and the practical point of view, this was not a sufficient reason to motivate a younger generation of scholars to follow in the footsteps of their elders. Indeed, in the early 1980s, the reasons to look at these rural communities appeared far less compelling than they had ten or twenty years earlier. As a matter of fact, even twenty years earlier, some sociologists, such as Ben-David, had felt that the moshav and the kibbutz did not deserve to occupy the attention of almost half of the Israeli sociological community, not being by themselves as nationally important or theoretically interesting as all other aspects of Israeli society put together.

### The Shifting Foci of Interest in Sociology and Anthropology

In the 1970s, the moshav became less interesting for those inclined to fund problem-oriented research. It was not a "locus of strain" (Handelman and Deshen 1975) as much as it had been before: Large numbers of moshavim seemed to have improved their lot and come out of trouble. The 1970s were a period of prosperity for all of Israel's agricultural sector. Indeed, Gluckman's statement that things were going well "except in a few ecological pockets" (Shokeid 1985) was not devoid of a factual basis.

In the meanwhile, the foci of interest in anthropology and sociology had shifted. One of the shifts was from rural to urban settings, to some extent because of the funds available for studying the latter thanks to the beginning of "Project Renewal" in the development towns and the urban neighborhoods launched by the Begin cabinet (Hazan 1988; Carmon 1989; Kressel 1993; Spiro 1988). Some of that shift was between the recipients of research budgets among the institutions of higher learning. Thus the Technion, always more oriented to the urban than to the rural scene, increased its share of the social research money allocated by the Jewish Agency.

Another shift specific to anthropology was from concrete to symbolic issues. Community studies became less frequent, not only because the communities were less and less insulated from the rest of the society but also because of the changing fashions of social science. The new trends may have been enhanced by the attractiveness of "armchair anthropology" and by the reluctance to undergo the inconvenience of traveling to faraway places and to be separated from the amenities of big city life. Indeed, in the late 1970s, few community studies were published and even finding the community within the confines of the American big cities (Suttles 1972) or discovering symbolic communities (Hunter 1976) seems to have gone out of fashion. Indeed, in Siegel's recent review of anthropology, the term community study is not even mentioned, and the only communities mentioned in the bibliography are the "imaginary communities" of nationalism (Siegel 1993). In addition to this, a large proportion of the moshavim (especially among those situated close to the big cities or in the center of the country, such as the moshav described by Kressel in this book) have become "commuter communities" themselves, participating in the process of urbanization and gentrification of Israel's countryside (Applebaum 1986; Schwartz and Applebaum 1994), and some kibbutzim may well be on their way to "jump on the bandwagon."

During the years of Likud government, the government made a large effort to set up many community settlements in the West Bank and in the Galilee and to spread them as much as possible. The community settlements are nonagricultural villages that have served the dual purpose of offering the urban middle class the opportunity to live in the country with a red tile roof and some grass near their houses and of providing the state with the ability to ensure Jewish presence in regions where it was felt that the demographic preeminence of Jews was threatened (e.g., the Galilee) or where there was a political will to set up Jewish implantations (e.g., Judea and Samaria—the West Bank). Most community settlements were set up in mountain regions, which in any case were devoid of agricultural possibilities. While most community villages were for all practical purposes commuter communities, some of them did invest considerable effort in setting up at least some businesses and economic branches (Newman 1984; Applebaum and Newman 1988). In the public opinion of Israel, the community settlements are associated with the ultranationalist Gush Emunim movement.

This effort involved not only budget shifts from the old type of cooperative and collective settlements (kibbutzim and moshavim) to the new type (commuter communities), but also an attempt to replace the kibbutz and the moshav as national myths by the settlements of the West Bank. Moshavim and kibbutzim started to be seen by many Israeli almost as anachronisms. Overall, the rural sector had lost much of its national importance since the heroic period of state-building.

In the 1980s, institutional support for kibbutz, and especially for moshav studies, started to grow thin. The budgets of the Settlement Department were progressively eroded, and the resources of other institutions which had traditionally funded research such as the kibbutz movements were also depleted, especially after the onset of the financial crisis. Thus, financial support for the study of the moshav and the kibbutz has lessened, while other subjects receive better funding.

These were, so it would seem, rather powerful reasons for young scholars to choose more important objects of study and objects more ensured of a stable future. The only new research undertaken in the early 1980s about the kibbutz was being pursued at the research institutes of the kibbutz movements, while the Settlement Studies Center had the practical monopoly of research about the moshav. Several moshav studies were started in the late 1970s or early 1980s: the study by Sherman and Banin (1992) about the attitudes of the settlers toward their regional organizations, the study by Arnon and Margolis (1984) about planning and reality in the Lakhish Region (moshavim set up in the mid-1950s), the study of Sherman, Sternberg, and Schwartz (1984) about the moshavim of the Jordan Valley, the study about the employment problems of moshav youth in the Galilee (Schwartz, Finkel, and Bar-El 1982), and also the study about the efficiency problems of the moshav association (Margolis and Cahan 1981).

What characterized these studies as opposed to the moshav studies of the past was first of all the populations they dealt with. The settlers of the new moshavim of the Jordan Valley were quite different from those of the moshavim of the 1950s, and, among the old clientele of the Settlement Department, research was mostly about the mountain moshavim, especially those of the Galilee, which had not succeeded in taking advantage of the financial assistance of the Department.

The new research paid special attention to the wider environment and to its transformations, including the economy (inflation, government subsidies, interest rates), political environment (consensus about the territories, consequences of the access to power of a Likud government), and the organizational environment (the weakening of the Settlement Department, the interplay between it and various other organizations). These environmental elements impinged more and more obviously upon what was happening in the moshavim as well as other settlements. Also, the new research was more critical than the previous research.

## THE CRISIS OF THE MID-1980S AND ITS AFTERMATH

At the Settlement Studies Center, there were some new research activities, especially in 1987 and after (e.g., after the onset of the severe financial crisis which has strongly affected the rural sector ever since). Much of that new research was a direct outcome of the crisis in the settlement sector, including the moshavim. Thus, among the eight research reports published in 1988, two were directly concerned with the moshav (Schwartz, Marvid, and Arian 1988; Shoresh 1988), the first with the causes and consequences of the financial crisis of the moshavim and the second with long-term process of differentiation which contributed to the decline in the effectiveness of cooperative functioning. A third project (Haruvi, Fleischer, and Eger 1989) dealt with the disappearance of a separate moshav labor market, a process speeded up by the moshav crisis.

Much of the research in 1989 and 1990 was also intended to improve coping with the results of the crisis (Schwartz, Sherman, Heller, and Eyal 1989; Applebaum, Shoresh, Eger, and Hartman 1990), the first to find out whether the regional councils could from now on provide the services which many moshavim had become incapable of providing, and the second to find out what forms of organization were now available to the moshavim. Other topics of research asked the question of how able the moshav second generation was to adapt to the new post-crisis situation, especially in the peripheral regions such as the Galilee where even before the crisis the situation was far from ideal. A perennial question was that of the stability of the moshav population in such regions and of what could be done to help it.

The Settlement Studies Center was well aware that the funding for moshav studies, and more generally, settlement studies, would diminish in the future due to the expected shrinking and disappearance of the Settlement Department, also one of the results of the settlement sector crisis. Thus, the center changed its name in 1989 to the Development Study Center after pursuing a number of projects about the moshav which were a direct outcome of the crisis.

In addition to the Development Study Center, some research and interest in the moshav also continued to exist among the personnel of the Settlement Department, sociologists and planners, some of whom did interesting pieces of research about moshavim (though much of their time was spent studying community settlements). Examples are the work of Nevo and Solomonica (1983) about the status of women in the moshav and the work of Nevo, Solomonica, and Peretz (1983) about the election results in the moshav sector.

Finally, some of the anthropologists who had pioneered the study of the moshav maintained an active interest in it. The one whose interest in the moshav was the steadiest was Shokeid, who published a number of papers, most of them about Romema, the community where he had done his doctoral field work.

In parallel to the revival of interest in the moshav, there also has been a considerable revival of interest in the kibbutz. Some books have been published about the crisis of the kibbutz and its aftermath, and a considerable research effort has been undertaken by the research institutes of the kibbutzim. The team at Yad Tabenkin, which is the research institute of the United Kibbutz Movement, Takam, the largest kibbutz federation in Israel, includes both sociologists and anthropologists (Ben-Rafael, Toppel, Sheffer, Rozolio) and economists (Don, Kislev) and has already published a considerable number of reports (Abrahami 1993).

## REFERENCES

Abrahami, E. 1993. *The Kibbutz as a Political System—The Loss of Will and of Standing*. Ramat Ephal: Yad Tabenkin.

Antonovsky, A., and A. Arian. 1972. *Hopes and Fears of Israelis: Consensus in a New Society*. Jerusalem: Academic Press.

Applebaum, L. 1986. "Migration of Urban Families to the Moshav in Israel 1968–1978." Ph.D. diss. Jerusalem: Hebrew University.

Applebaum, L., and J. Margolis. 1979. *The Moshav and the Test of Time—Patterns of Change in an Organization*. Rehovot: Settlement Studies Center (Hebrew).

Applebaum, L., and D. Newman. 1988. *Settlement Villages in the West Bank*. Rehovot and Beer Sheva: Settlement Studies Center and Ben-Gurion University Press.

Applebaum, L., D. Shoresh, M. Eger, and S. Hartman. 1990. Alternatives to Association in the Moshav. Research report. Rehovot: Development Study Center.

Arensberg, C. M. 1937. *The Irish Countryman*. Cambridge, Mass.: Harvard University Press.

———. 1954. The Community Study Method. *American Journal of Sociology* 60: 109-124.

Arnon, I., and J. Margolis. 1984. The Planning of the Moshav and the Test of Reality: The Transformations of the Economy in the Lakhish Region. Working paper. Rehovot: Settlement Studies Center (Publications in Regional Development 35 [Hebrew]).

Baldwin, E. 1972. *Differentiation and Cooperation in an Israeli Veteran Moshav*. Manchester: Manchester University Press.

Barkai, H. 1977. *Growth Patterns of the Kibbutz Economy*. Amsterdam: North Holland.

Ben-David, J. 1964. "The Kibbutz and the Moshav." In *Agricultural Planning and Rural Community in Israel*, J. Ben-David, ed. Paris: UNESCO. (Arid Zone Research 23.) Reprinted in *The Sociology of the Kibbutz*, E. Krausz, ed. New Brunswick, N.J.: Transaction Books.

———. 1972. Personal communication.

Ben-Rafael, E. 1980. "Dynamics of Social Stratification in Kibbutzim." *International Journal of Comparative Sociology* 21(1–2): 88–100.

Bott, E. 1957. *Family and Social Network*. London: Tavistock.

Buber, M. 1949. "An Experiment That Did Not Fail." In *Paths in Utopia*, pp. 139–149. Boston: Beacon Press.

Carmon, N. 1989. *Neighborhood Rehabilitation in Israel: Evaluation of Outcomes*. Faculty of Agriculture and Town Planning and the Samuel Neaman Institute of Advanced Studies in Science and Technology. Haifa: The Technion, Israel Institute of Technology.

Cohen, E. 1976. "The Social Transformation of the Kibbutz." In *Social Change*, K. Zollschan and W. Hirsch, eds. New York: John Wiley.

———. 1982. "The Black Panthers and Israeli Society." *Jewish Journal of Sociology* 14(1): 93–109.

Diamond, S. 1957a. "Kibbutz and Shtetl: The History of an Idea." *Social Problems* 5: 71–79.

———. 1957b. "Kibbutz: Utopia in Crisis." *Dissent* 4: 40–46.

Don, Y., B. Ilan, R. Bar-El, A. Bar Hai, I. Greenspan, and D. Schnitt. 1977. *Moshav Industrialization*. Rehovot: Settlement Studies Center.

Ellen, R. F., ed. 1984. *Ethnographic Research: A Guide to Ethnographic Research*. Orlando, Fla.: Academic Press.

Goldberg, H. 1972. *Cave Dwellers and Citrus Growers: A Jewish Community in Libya and Israel*. Cambridge: Cambridge University Press.

Handelman, D., and S. Deshen. 1975. *The Social Anthropology of Israel: A Bibliographical Essay*. Tel Aviv: Tel Aviv University.

Haruvi, N., A. Fleischer, and M. Eger. 1989. The Employment Structure in the Moshav Sector. Research report. Rehovot: Center for Development Study.

Hazan, H. 1988. "Dora—A Renewal Neighborhood as a Community—An Anthropological Analysis from a Social World Perspective." *Megamot. Neighborhood Renewal in Israel* (3–4): 286–299 (Hebrew).

Helman, A. 1985. "The Manager Has Been a Manager and Shall Remain a Manager." *Yahad* (The Organ of the United Kibbutz Federation [Hebrew]).

———. 1987. "Professional Managers in the Kibbutz." *Riv'on Lekalkala (Economic Quarterly)* 131: 1031–1039 (Hebrew).

Hunter, A. 1976. *Symbolic Communities. The Persistence and Change of Chicago's Local Communities*. Chicago: University of Chicago Press.

Krausz, E., ed. 1983. *The Sociology of the Kibbutz*. Vol. 2, *Studies of Israeli Society Series*, The Israeli Sociological Society. New Brunswick, N.J.: Transaction Books.

Kressel, G. 1974. *From Each According to His Ability: Stratification vs. Equality in a Community in Transition from Agriculture to Industry.* Tel Aviv: Tchirikover (Hebrew).

———. 1983. *To Each According to His Needs: Equity vs. Equality in the Allocation of Rewards in a Community in Transition from Agriculture to Industry.* Tel Aviv: Tchirikover (Hebrew).

———. 1993. "Who Is Fed Up Working and Why; Unemployment in Ashkelon." *Studies in the Revival of Israel* 3: 168–190 (Hebrew).

Kushner, G. 1973. *Immigrants from India in Israel: Planned Change in an Administered Community.* Tucson: University of Arizona Press.

Landshut, S. 1944. *The Kvutza: A Sociological Study of the Kibbutz Type Settlement in E Eretz Israel.* Jerusalem: The Small Zionist Library, The Institute for Zionist Education (Hebrew).

Leviathan, U., and M. Rosner, eds. 1980. *Work and Organization in the Kibbutz Industry.* Norwood, Pa.: Norwood Editions.

Liebow, E. 1975. *Tally's Corner: A Study of Negro Streetcorner Men.* London: Routledge and K. Paul.

Margolis, J., and D. Cahan. 1981. Effectiveness and Efficiency of Moshav Associations. A Summary of Findings and Recommendations. Research report. Rehovot: Settlement Studies Center.

Meshorer, R. 1987. *Bibliography of the Moshav.* Rehovot: Settlement Studies Center.

Nash, R. 1962. *Machine Age Maya.* Chicago: University of Chicago Press.

Nevo, N. 1978. "Social Change in a Veteran Moshav Ovdim." Ph.D. diss. Tel Aviv: Tel Aviv University.

———. 1993a. "The Anthropologist in a Rural Bureaucracy: Chances for Survival." *Practicing Anthropology* 5: 9–12.

———. 1993b. Personal communication.

Nevo, N., and D. Solomonica. 1983. *Ideological Change of Rural Women's Role and Status: Case Study of Family Based Cooperative Villages in Israel.* Ypsilante: Michigan State University, Work and Development Services.

Nevo, N., D. Solomonica, and H. Peretz. 1983. *Voting Patterns of Moshavim in the 1981 Elections.* The Jewish Agency and the World Zionist Organization, Rural Settlement Department and Rural Settlement Division (Hebrew).

Newman, D. 1984. "The Development of the Yshuv Kehilati in Judea and Samaria: Political Process and Settlement Form." *Tijdschrift voor Economische and Soziale Geographie* 75: 140–150.

Redfield, R. 1959. *The Little Community.* Chicago: University of Chicago Press.

Rosenfeld, E. 1951. "Social Stratification in a 'Classless' Society." *American Sociological Review* 16: 766–774.

———. 1958. "The American Social Scientist in Israel: A Case Study in Role Conflict." *American Journal of Orthopsychiatry* 28: 563–571.

Rosner, M. 1972. "Women in the Kibbutz." In *Readings in the Psychology of Women*, J. Bardwick, ed. New York: Harper & Row.

Rosner, M., J. Ben-David, A. Avnat, N. Cohen, and U. Leviathan. 1990. *The Second Generation: Continuity and Change in the Kibbutz.* Westport,

Conn.: Greenwood Press (Hebrew edition published in 1978 by Sifriat Hapoalim and the Settlement Studies Center).

Rosner, M., and N. Cohen. 1980. "Is Direct Democracy Feasible in Modern Society? The Lesson of the Kibbutz Experience." In *Integrated Co-operatives in the Industrial Society: The Example of the Kibbutz*, K. Bartolke, T. Bergman, and L. Liegle, eds. Assen: Van Gorcum.

Schwartz, M. 1995. *Unlimited Guarantees: History and Political Economy in Israel's Cooperative Agriculture*. Beer Sheva: Ben-Gurion University Press (Hebrew).

Schwartz, M., and L. Applebaum. 1994. *The Regional Councils in a Changing Environment: The Challenge of Development Planning*. Rehovot: Development Study Center.

Schwartz, M., R. Finkel, and R. Bar-El. 1982. Non-Agricultural Employment for Moshav Youth in the Galilee. Research report. Rehovot: Settlement Studies Center.

Schwartz, M., R. Finkel, R. Bar-El, and A. Nesher. 1986. "Moshav Based Industry." Reprinted in *Rural Industrialization in Israel*, R. Bar-El, ed., 1988. New York: Westview.

Schwartz, M., and D. Giladi. 1993. "25% Farmers? Plans and Reality in the First Decade of the State of Israel." *Riv'on Lekalkala (Economic Quarterly)* November: 391–414 (Hebrew).

Schwartz, M., A. Marvid, and O. Arian. 1988. *The Organizations: Three Moshav Purchasing Cooperatives before and after the Crisis*. 2 Vols. Rehovot: Settlement Studies Center (Hebrew).

Schwartz, M., N. Sherman, D. Heller, and A. Eyal. 1989. Education and Municipal Services in Three Moshav Regional Councils. Research report. Rehovot: Development Study Center (Hebrew).

Schwartz, R. D. 1954. "Social Factors in the Development of Legal Control: A Sociological Analysis of the Israeli Kvutza." *Yale Law Journal* 63: 471–493.

———. 1955. "Functional Alternatives to Inequality." *American Sociological Review* 20: 424–430.

Shapiro, O., ed. 1971. *Rural Settlements of New Immigrants in Israel*. Rehovot: Settlement Studies Center.

Shepher, I. 1983. *The Kibbutz—An Anthropological Study*. Norwood, Pa.: Norwood Editions.

Shepher, J. 1970. "Organizational Effectiveness in Kibbutz Society in the Light of New Theory." *Sociologia Ruralis* 10(1): 212–237.

Sherman, N., and T. Banin. 1992. *The Attitudes of Moshav Members Regarding the National Moshav Movements: Findings of a Public Opinion Survey*. Rehovot: Development Study Center (Hebrew).

Sherman N., M. Schwartz, and E. Sternberg. 1984. "The Guided Moshavim Method and Its Implementation in the Moshavim of the Jordan Valley." In *Human Settlement and Regional Development*, A. Rosenman, ed. Jerusalem: Keter.

Shokeid, M. 1985. *The Dual Heritage: Immigrants from the Atlas Mountains in an Israeli Village*. New Brunswick, N.J.: Transaction Books (first printed in 1971 by Manchester University Press).

Shoresh, D. 1988. Farm Differentiation in the Moshav. Working Paper 19 (Hebrew). Rehovot: Development Study Center.

Siegel, B. J. 1993. "The First Twenty Years." *Annual Review of Anthropology* 22: 1–34.

Spiro, M. E. 1954. "Is the Family Universal?" *American Anthropologist* 56: 839–846.

———. 1956. *Venture in Utopia.* Cambridge, Mass.: Harvard University Press.

Spiro, S. E. 1988. "Issues in the Evaluation of Neighborhood Rehabilitation Programs: Lessons from Israel's Project Renewal." *Megamot: Neighborhood Renewal in Israel* (3–4): 269-286 (Hebrew).

Stone, P. 1972. Personal communication.

Suttles, G. D. 1972. *The Social Construction of Communities.* Chicago: University of Chicago Press.

Talmon-Garber, Y. 1972. *Family and Community in the Kibbutz.* Cambridge, Mass.: Harvard University Press.

Tiger, L., and J. Shepher. 1975. *Women in the Kibbutz.* New York: Harcourt Brace.

Warner, W. L., and P. Lunt. 1941. *The Social Life of a Modern Community.* Yankee City Series, Vol. 1. New Haven: Yale University Press.

Weingrod, A. 1966. *Reluctant Pioneers: Village Development in Israel.* Ithaca: Cornell University Press.

———. 1993. Personal communication.

———. 1994. Personal communication.

Weintraub, D., and F. Bernstein. 1966. "Social Structure and Modernization, a Comparative Study of Two Villages." *American Journal of Sociology* 71(5): 509–521.

Weitz, R. 1962. "The Role of the Sociologist in Israel's Rural Development." *Transactions of the Fifth World Congress of Sociology,* Vol. 1. Reprinted in *Rural Settlements of New Immigrants in Israel,* O. Shapiro, ed. pp. 15–28. Rehovot: Settlement Studies Center.

Weller, L. 1974. *Sociology in Israel.* Westport, Conn.: Greenwood Press.

Whyte, W. F. 1943. *Street Corner Society: The Social Structure of an Italian Slum.* Chicago: University of Chicago Press.

Willner, D. 1969. *Nation-Building and Community in Israel.* Princeton: Princeton University Press.

Yuchtman-Yaar, E. 1972. "Reward Distribution and Work Role Attractiveness in the Kibbutz: Reflections on Equity." *American Sociological Review* 35(7): 581–595.

———. 1983. "The Tension between Communal Values and Economic Imperatives in the Modern Kibbutz." *Contemporary Sociology* 12(5): 509–512.

Zborowski, M., and E. Herzog. 1952. *Life Is with People.* New York: International Universities Press.

Zussman, P. 1988. *Individual Behavior and Social Choice in a Cooperative Setting: The Theory and Practice of the Israeli Moshav.* Jerusalem: Magnes Press.

## Chapter 2

# The Effects of Public Financial Assistance on the Management of Moshav Economic Affairs

*Neal Sherman and Moshe Schwartz*

*This chapter analyzes the impact of public financial support on economic behavior in* moshavim. *Until the recent (1986 to 1990) crisis of the rural settlement sector, the public institutions responsible for settlement were unwilling to allow a* moshav *to reach a state of economic breakdown that might lead to its abandonment. If debts reached a critical level, there was a high probability that emergency assistance would be provided. Examination of events in moshavim established in the Jordan Valley after 1967 shows that the expectation of aid from public agencies weakened the pressure felt by members and management to conduct the cooperative's affairs in an efficient, orderly, and disciplined manner. Similarly, since the settlers' financial obligations were likely to be covered by an influx of public funds, one member was not likely to see in another's debt a direct financial burden. Demands directed to heavy debtors were thus likely to be interpreted as unreasonable meanness toward those who faced the same difficulties and hardships shared by all. The expectation of assistance undermined the moshav's ability to impose financial discipline on individual members.*

Examination of the economic financial behavior of the moshavim established in the Jordan Valley after 1967 shows that public funding and its expected continuation undermined cooperative discipline and both encouraged and made possible the irresponsible management of moshav operations.

The moshav is a small holders' cooperative village based on the principle of "mutual liability"; that is, members are expected to sign financial guarantees in favor of the moshav in which they assume liability for the full extent of the cooperative's indebtedness. Cooperative theory suggests that the members' interest in maximizing the return from their participation in the cooperative and their formal responsibility for the cooperative's financial obligations will lead them to attempt to ensure the prudent management of cooperative operations. In order to achieve this objective, members will exercise active oversight over management performance through the democratic structures of the society. They also will be concerned that all members maintain cooperative discipline and fulfill their obligations to the society. Thus, the members will take part in the application of social controls in order to counter deviant behavior and will also back officials' attempts to impose formal discipline.

The research on which this chapter is based demonstrates that the expectation that public financial assistance will be readily available to the moshav severely weakens the members' motivation to impose controls on cooperative officials and upon other members. Where it is felt that the bill for inefficiency and a lack of discipline can be passed on to the public treasury, members will refrain from entering into conflict with management and with each other in order to remedy the weaknesses of cooperative operations (Sherman and Schwartz 1987; Sherman, Schwartz, and Sternberg 1984; Sherman, Sternberg, Schwartz, and Perlman 1984; Sherman, Sternberg, Eger, and Cohen 1987). While public financial backing has been referred to in previous moshav studies, the consequences of such assistance have not yet been subject to extensive scrutiny, nor have they been the focus of an analysis of moshav-type settlements. The original impetus for the study of the moshav was its role in the absorption of new immigrants in Israeli society and culture. The process of sociocultural adaptation was thus the central research concern. Researchers analyzed the relations between the settlement organizations and the new immigrants as one of the factors affecting such adaptation. Some scholars also presented the guiding principles and administrative arrangements for the allocation of financial and other resources to the settlers by the settlement authorities. Conflicts within the village over control of such resources and the process by which the settlers learned to manipulate the system in order to press their demands also received attention. However, relatively little analysis was devoted to the impact of institutional economic support upon the detailed, day-to-day financial and economic management of individual farm units and the village cooperative (Shokeid 1971; Weingrod 1966; Weintraub 1971; Willner 1969).

The few available studies of veteran moshavim whose members had already passed the stage of adaptation to the moshav and Israeli society were also carried out by sociologists and social anthropologists whose professional background tended to limit their interest in the treatment of economic and financial matters (Abarbanel 1974; Baldwin 1972; Nevo 1978; Weintraub, Lissak, and Atzmon 1969). On the other hand, the economists who have studied the moshav have usually focused their interest on the agricultural productivity of individual small holdings rather than inquiring into the effect of financial support on farm management decisions (Haruvi and Kislev 1984; Mundlak 1964; Sadan and Weintraub 1980).

The findings summarized in the following discussion are derived from evaluation studies conducted in order to assess the effects of public financial assistance and the effectiveness of administrative intervention in moshav affairs which accompanied the provision of funds. In keeping with these objectives, research teams were recruited which included an expert in moshav accounting systems and financial analysis, in addition to sociologists. These evaluation studies thus concentrated on topics that were not of primary concern in previous work on the moshav.

The Israeli case study presented here is illustrative of internationally relevant policy issues. In the developing nations, policy makers have identified the cooperative as an appropriate means of carrying out a number of important rural and agricultural development functions. The natural conclusion there, as in Israel, has been that given its potential contribution, the cooperative should be given extensive government backing (Young, Sherman, and Rose 1981). Therefore, in the developing countries, as in Israel, an analysis of cooperative operations cannot be based solely on an examination of factors internal to the local community but must assign adequate weight to government intervention and its influence (Gyllstrom and Rundquist 1989). Many studies of developing country cooperatives have emphasized that heavy handed administrative intervention can alienate farmers from the cooperative by negating its character as a democratically managed community organization (Gyllstrom and Rundquist 1989; Hyden 1984). This chapter's analysis of the Israeli experience demonstrates that too open-handed an approach can be harmful as well.

The reader is requested to note that the findings presented in this chapter depict the nature of settlement authority relations during the years until the collapse of the regional purchasing cooperatives in 1985 and 1986. The failure of public institutions to step in and provide the assistance needed to save these cooperatives from bankruptcy marks a watershed in the history of the rural settlement sector and its relations with Israeli society (Schwartz, Marvid, and Arian 1988; Sherman 1993).

Prior to the recent crisis, a balance had prevailed between the special status of the cooperative settlement sector and the extent of the sector's need for direct public financial support and informal, implicit loan guarantees. By 1985–1986, the sector's public standing had been eroded because of changes in the prevalent values of Israeli society and because of the declining importance of rural cooperative settlement as means to the realization of national goals. In the years leading up to 1985, the settlement cooperatives' debts had grown to unprecedented proportions. By the time the purchasing cooperatives asked for help, the need for financial assistance far outstripped the government's willingness to pay.

It is not yet clear what the detailed character of the new system of relations between the cooperatives and public institutions will be, but it is already obvious that the priority assigned to rural settlement cooperatives in the allocation of public financial resources will not be restored to its previous high level. While a number of generalizations presented in this chapter are phrased in the present tense, it should be kept in mind that the system described is currently undergoing a series of changes that may be expected to alter its most basic characteristics.

## THE MOSHAVIM AND THEIR MEMBERS

The roots of the Israeli system of public financial backing for new rural settlement are to be found in the symbolic and practical roles that the settlers of the working (i.e., moshav cooperative and kibbutz communal) settlement sector played in the Jewish community of preindependence Israel. The members of moshavim and kibbutzim came to be seen as valiant pioneers, serving the Jewish people by carrying out the practical settlement objectives of the Zionist movement (Sherman 1980). Espousal of the principles of nationally owned land and self-labor by the members of the moshavim and kibbutzim helped them to monopolize the symbolic value adhering in Zionist ideology to the return to the land through agriculture (Shapira 1973). It was generally accepted that worker-settlers could not accumulate sufficient capital of their own to establish themselves as farmers. Therefore, the harnessing of their enthusiasm and idealism to the benefit of Zionist settlement goals demanded that the institutions of the Zionist movement make available to the worker-pioneers the financial assistance they required. In this relationship, the settlers were cast in the value-laden role of pioneers, while the staff of the supporting organizations were cast as bureaucrats whose role was legitimized by the need to support the pioneers (Sherman 1979).

These generalizations regarding the settlers' advantage over the bureaucrat are most valid with regard to the preindependence period and the post-1967 period. During the immediate postindependence years of mass immigration and settlement, the new immigrants sent to rural settlement did not, in the great majority of cases, hold the social and political values that the pioneer was presumed to espouse (Weingrod 1966). Indeed, many of the new settlers from North Africa and the Middle East were seen as representing a "primitive," premodern culture and thus were viewed as requiring guidance and cultural formation (Kushner 1973; Willner 1969). In this context, the instructors were the actors more strongly imbued with pioneering, and, in general, they established Israeli cultural conceptions and values. In the new immigrant settlements, the balance of power in relations between the settlement institutions and the settlers swung about to the advantage of the institutions.

As the new immigrant-settlers became more thoroughly acculturated, they came to understand the limits of bureaucratic power in an open, democratic, party-political system and were able to revise the character of relations with the settlement authorities in their favor (Weingrod 1962; Willner 1969). During the 1960s, the ability to influence settler behavior through allocation or denial of economic resources was gradually transferred from public bureaucracies to the hands of leaders who were drawn from among the immigrants themselves and were recruited to positions in political parties and in the politically oriented settlement movements with which the individual moshavim are affiliated.

The leadership group of the new moshavim gave a first major demonstration of its power when it succeeded in circumventing the attempt to impose discipline on the settlements through the centralized credit system. The centralized credit system was initiated by the Ministry of Agriculture in 1962. According to its regulations, the allocation of various types of subsidized credit was made dependent upon the channeling of all moshav transactions through a single bank that was to regulate all credit, both subsidized and commercial, in keeping with sound business criteria (Abarbanel 1974; Gvati 1981). The heads of the new immigrant moshavim were, however, able to use the regional purchasing cooperatives in which their settlements were organized in order to neutralize financial control by the banks. Large turnover and political influence enabled the purchasing cooperatives to mobilize credit which they then made available to member moshavim. The purchasing cooperatives, controlled by movement politicians, thus became the real bank for most of the moshavim established in the postindependence period. The development of this independent source of credit strengthened significantly the autonomy of the new immigrant settlement sector vis-à-vis the settlement authorities (Schwartz, Bar-El, Finkel, and Nesher 1987).

A further decisive demonstration of the political strength and maturity of the new immigrant moshavim came in 1966 when their backing served as the key factor protecting the Settlement Department of the Jewish Agency against those advocating a severe cutback in its resources and autonomy, or even its dissolution (see the discussion in the following section). The role that the new immigrant moshavim played in this political struggle marked their accession to a position of influence similar to that of the veteran settlements in their relations with the public authorities responsible for agriculture and rural settlement (Schwartz, Marvid, and Arian 1988).

The year 1967 marked the beginning of a period of new settlement activity concentrated in areas that had come under Israeli control during the Six Day War. For a number of years, there was a revival of both settlement activity and popular enthusiasm regarding the settlement effort. New settlements continued to benefit from the supports institutionalized in the preindependence period, but it became clear that the balance of elements composing the Israeli value system had shifted, resulting in a change in the evaluation of new settlement. While settlement was still seen as a collective goal, both the general public and the settlers, primarily native Israelis, began to feel that the desire to further national and social ideals was no longer a sufficient motive to bring candidates to a new settlement point. The new settlement should offer advantageous conditions for the realization of personal aspirations: It would do so primarily by allowing members to achieve a comfortable standard of living. While the settlers were still in some sense viewed as pioneers, they could hardly succeed in assuming the aura of selflessness that enveloped their preindependence predecessors. It was clear both to the settlers and to the public that they would not be making the contribution they were making if there were not something concrete in it for them personally.

## THE SETTLEMENT DEPARTMENT AND THE STRUCTURE OF SETTLER–SETTLEMENT AGENCY RELATIONS

The principle of public backing for the establishment of cooperatively based settlements was formalized and codified in plan estimates and later in a "settlement key" published in 1929 (Preparatory Commissions 1929). Taking the provision of a reasonable, modest standard of living as their basic objective, the authors of these documents derived the investment required to establish a family farm unit in the various ecological zones of the country. Once formulated, the settlement key served as the authoritative definition of the settlers' entitlement to receive investment capital from the authorities (Rokach 1986).

The Settlement Department (Division)[1] is the organizational unit directly responsible for administering the special settlement budget allocated to new villages according to the provisions of the settlement key. The Department also provides a variety of technical services to new settlements and has a diffuse, general responsibility for oversight and guidance of the establishment and breaking in of new villages, particularly with regard to economic and financial aspects.

Debt relief measures are a logical derivative of the planned, highly subsidized provision of land and capital for the establishment of new settlements. The foundation of settlement activities on plans aimed at enabling the settler to make a living from his farm unit; the gradual provision of investment funds within the framework of the settlement key rather than in a lump sum; and the provision of technical services aimed at making possible the realization of planning goals—these factors together constitute an implicit guarantee that operation of a farm, as specified in the settlement plans, will make it possible for the settlers to earn an adequate livelihood. The Settlement Department's involvement in planning, investment, oversight, and guidance substantiate a claim often raised by settlers facing economic difficulties: The Department must come to their aid and provide special assistance in order to allow them to maintain a reasonable standard of living.

Practical political considerations ensure that the Settlement Department will take concrete steps in keeping with its public, moral responsibility for the settlers' economic situation. The Department is sensitive to the fact that, given its central role in settlement planning and implementation, the failure and abandonment of a settlement would inevitably be interpreted by the general public as a direct failure on the part of the Department itself, whatever the responsibility of the settlers themselves might be. Furthermore, the Department is aware that in any public argument between it and the settlers as to the apportioning of the blame for settlement failures, it would almost inevitably emerge the loser. As mentioned above, the settlers fill the value-laden role of pioneers, while the staff of the Settlement Department is cast in the role of bureaucrats. The moral value of the bureaucratic role is derived from the value of the pioneering role; the settlement bureaucrat exists to serve the pioneer. Indeed, the Department's very right to exist as an independent agency in the postindependence period after the establishment of the state Ministry of Agriculture is derived from the special charisma of the pioneer and the act of new land settlement. Finally, as indicated earlier, the moral symbolic superiority of the settler is reinforced by the political dependence of the Department on the backing of the nationwide settlement movements and the regional economic organizations with which the moshavim are affiliated. These bodies constitute the organized clientele that provides to the Department the political support needed to protect its autonomy and advance its demands for budgetary resources.

The extension of the Settlement Department's role in relation to the new settlements, from the provision of basic investment capital to the provision of emergency debt relief, is thus an outgrowth of the Department's responsibility for investment in the establishment of new settlements, reinforced by political necessity. For the reasons just detailed, the Department cannot stand by and see a moshav reach a state of economic breakdown that would lead to its abandonment by its settler members. This basic principle was found to lie at the core of both settler's and staff members' understanding of the political, as well as the financial, relations between the moshavim and the Department.

Investment, the proclaimed core activity of the Department, is formally bounded by the settlement key, but the extent of ongoing support of various kinds and, in particular, the precise circumstances justifying the provision of debt-relief assistance and the ultimate magnitude of such assistance are much less clearly established. The absence of clear, formal regulations for the provision of emergency assistance has persisted, despite the fact that the rescue of settlements from burdensome debts has been an important aspect of new settlement endeavors since the earliest days of modern Zionist activity in the land of Israel (Ettinger 1945). Technical difficulties in the precise determination of the causes of economic and financial difficulties account for something of this chronic inability to fix the boundaries between the settlement authorities' responsibility and that of the settlers themselves. The main cause of the inability to define the limits of special financial assistance would seem to be rooted in the ideological and practical political considerations outlined previously.

In such a situation, there is a real possibility that emergency debt relief will be converted, in practice, to recurrent, periodic assistance. The Department's aid cancels for the settlers the effects of market discipline and the discipline imposed on clients by commercial financial institutions. Since the Department lacks the moral and political strength to withhold aid consistently from miscreant and incompetent settlers and settlements, its direction and oversight cannot replace market discipline. Public support, channeled through the Settlement Department, makes responsible behavior dependent on the existence of particularly strong individual values on the one hand and community values and effective social controls on the other.

As noted earlier, the Settlement Department enjoyed a more authoritative status vis-à-vis the settlements under its care during the period of mass settlement by new immigrants from the Middle East and North Africa. The Department was granted large budgets, and its personnel expanded to peak levels, with instructors present in each village. Furthermore, the mass character of the settlement enterprise during this period and the ready store of replacements for individuals and groups leaving the villages weakened the leverage that settlers and settlements could bring to bear on the Settlement Department.

As the studies focused on new immigrant moshavim show, however, even at the height of its relative advantage, the ability of the Settlement Department to control the ongoing management of family farms and moshav affairs had clear limitations. Basically, the Department's efforts to bring about a deep internalization of cooperative values were in most cases unsuccessful. Furthermore, despite the Department's close supervision, many settlements began to run up uncontrollable debts as settlers turned to nonagricultural pursuits and neglected their farms. Marketing of agricultural produce outside of cooperative channels also became more prevalent and contributed to the worsening financial situation of many new moshavim (Shokeid 1971; Weingrod 1966). As discussed earlier, the settlers' increasing understanding of Israeli culture and the Israeli political system, the rise of sectoral leadership from within the ranks of the settlers, and the development of the regional purchasing cooperatives as an autonomous base of economic and political power eventually freed the new immigrant moshavim from their position of weakness.

When public controversy arose in the mid-1960s concerning the status and role of the Settlement Department, the new immigrant settlements were able to show clearly that the tables had been turned in their relations with the Department. With the end of the years of mass immigration, the establishment of new settlements slowed and reached its lowest ebb during this period. The right of the Settlement Department to maintain its large staff and, indeed, its very existence as an autonomous administrative agency was increasingly questioned within the Israeli political system. This debate led in 1966 to the establishment of the Ben Aharon Committee to review the role of the Settlement Department. The political support of the leaders of the new-immigrant moshav sector proved to be of vital importance to the Settlement Department's campaign to avert the worst of the dangers represented by the establishment of the Committee. Nonetheless, the Committee's recommendations did lead to cuts in the Settlement Department's personnel and budgets and the transfer of a number of Department functions to the Ministry of Agriculture and other government bodies (Public Committee for Settlement Affairs 1966; Schwartz, Marvid, and Arian 1988).

The revival of settlement activity in the wake of the Six Day War gave the Settlement Department (and the Settlement Division, which acted in its stead in the occupied territories) a new lease on life. The Department had a vital interest in exploiting the return of new settlement to the headlines in order to enhance the public image of settlers and settlement upon which its own public standing was based. Although the behavior of the settlers often failed to fit the pioneering model, it was hard for the Department to call them to order publicly without damaging its own interests. The Department's influence over the new settlers was also limited by the fact that the great majority of them were native Israelis rather than new immigrants. Furthermore, the new settlements were affiliated with the settlement movements and purchasing cooperatives upon which the Department had become increasingly dependent for crucial political backing. The Department thus had a limited ability to exercise control over settlers and communities that ignored its advice and violated its directives.

The following sections illustrate the operation and impact of the system of special financial support for new settlement in an area whose values were no longer the same as those of the period in which the basic principles of public support were codified.

## THE CASE OF THE POST-1967 MOSHAVIM IN THE JORDAN VALLEY

The highest degree of protection against financial difficulties and possible bankruptcy is expected by settlements in regions that are sensitive from the point of view of defense considerations and government's overall stance regarding the Arab–Israeli territorial conflict. From both of these points of view, the moshavim established since 1969 in Biq'at Hayarden, the section of the Jordan Valley taken from Jordan in the Six Day War, fall into the category of those settlements entitled to the highest consideration.

The settlements in the Jordan Valley were a component in the Alon Plan, which laid out the Labor Party's basic stance regarding the West Bank. The objective of this plan was to ensure that the Jordan River would remain the security border of the State of Israel, while the densely populated territories in the heart of the West Bank were to be returned to Arab political control. The flight of most Arab residents of the Jordan Valley during the Six Day War and the period immediately following removed any demographic obstacle to the desired retention of the region under Israeli control; and according to prevalent security-settlement conceptions, civilian settlement was vital to the maintenance of the state's hold on territories to which it wished to assert a permanent claim (Harris 1980). Since settlement in the area was in keeping with the Labor Alignment's foreign and defense policy, the Jordan Valley became a desirable settlement objective for settlement movements associated with Labor, as well as for movements associated with religious and right-wing parties whose stance on the territorial issue demands retention of all of the territories taken in 1967.

Except for one settlement based primarily on new immigrants from France, the Jordan Valley moshavim are inhabited by Israeli-educated members, with only a sprinkling of immigrants who reached the moshavim as individuals. The ethnic origin of the population of the moshavim is mixed, with both the major cultural groupings, Ashkenazim and Sepharadim, represented in each settlement. Even though in a few of the settlements ethnic epithets were bandied about in the early years in moments of conflict and crisis, factional divisions did not, in fact, closely parallel ethnicity. During the later period covered by the two research projects that serve as the basis for this chapter, ethnic tensions were not a significant factor in the community life of any of the Jordan Valley moshavim.

The high priority assigned to Biq'at Hayarden and the consequent desire to move quickly in settling the area led paradoxically to steps that undermined the chances of successful development in most of the moshavim in the region. The majority of the Jordan Valley moshavim were set up by haphazardly assembled groups of individuals who did not share any substantial presettlement experiences. Organized presettlement activity initiated by the movements and the Settlement Division was either limited in scope and content or partial in coverage.

The desire to act quickly also meant the establishment of settlements based on agriculture before the accumulation of adequate data to make possible the identification of crops and crop varieties suitable to the subregions of the valley. In several settlements, many years passed before the settlers received their full complement of means of production. The public funding made available for important investment objects was inadequate in almost all the settlements, and, in many instances, it was paid over to the moshavim only after long delays. The settlers were forced to make use of high interest, short-term credit to make up the difference (the funding gap) between the actual investment outlay and the special investment credit received.

The guidance and oversight provided by the Settlement Division and the movements was, in general, quite inadequate. Little was done to prepare moshav members for management roles, either as elected committee members or as salaried officials of the moshav cooperative. The accounting systems of the moshavim were not given sufficient attention, and serious deficiencies in the financial data available compounded the managerial weaknesses of the settlements.

The changes in overall national economic and agricultural policy introduced by the right-wing Likud government beginning in 1977 had a devastating effect on most of the Jordan Valley moshavim established by that time and on those that were established after the Likud's rise to power (Sherman 1982). As the impacts of adverse policy measures and unfavorable economic conditions accumulated in the Jordan Valley and in the agricultural sector as a whole, the Ministry of Agriculture and the Settlement Division had to take a series of emergency debt relief measures. As a region of both high political priority and high debts, the Jordan Valley was among the prime recipients of such financial aid. The first payments were made on behalf of the moshavim to their respective purchasing cooperatives, in which the bulk of their hard debts were concentrated, in December 1979. The process of financial assistance then continued under a variety of guises until the end of the period covered by our field research in September 1984.

Soon after the provision of emergency aid commenced, the heads of the Division declared that they would not be satisfied with simply keeping the Biq'at Hayarden settlements afloat financially but rather intended to make use of the provision (or denial) of assistance as an incentive to more responsible financial management. Settlements managing their affairs in keeping with Division policy would receive the aid they needed; those violating the Division's expectations would risk the loss of its support.

To ensure that the moshavim fulfilled its requirements, the Division intensified its supervision of their financial affairs. The Division was no longer willing to limit itself to ex post facto evaluation of the implementation of annual economic plans. Instead, it attempted to impose a system of detailed, ongoing inspection of farming and financial operations. The Division's field staff made more frequent visits in each of the moshavim. During these visits, they physically inspected crops and observed the operations of moshav service facilities, such as packing houses, supply shops, and tractor and tractor implement pools. They met with the moshav's treasurer and its farm operations coordinator and heard reports on the performance of their duties. Finally, they inspected the moshav accounts, including a detailed review of check and receipt books.

On a more sophisticated level, an attempt was made to impose upon the moshavim a preplanned cash flow schedule. The regional purchasing cooperatives agreed to provide credit if the moshavim would deposit the income from all farm sales in their accounts in the organizations. The organizations were to release funds only according to the schedule authorized by the Division.

Through the application of such measures, the Division sought to bring about two major goals in the reform of moshav financial management: the distribution of deficits to members' private accounts and the regulation of their private consumption. The members of the cooperative society are fully responsible for all its debts, in keeping with the moshav principle of joint liability. The cooperative is therefore obliged to review its accounts periodically and debit the members with the full value of deficits.

The accumulation of very large undivided debts in the accounts of moshavim in Biq'at Hayarden was the result of several factors. First, undivided debts were the product of technical weakness in the accounting systems of the moshavim and undercharging for goods and services provided to members by the cooperatives. The failure to distribute deficits once their existence became known was a reflection of the weakness of cooperative authority and the fear of both elected and salaried officials regarding members' likely reaction to the debiting of large sums to their private accounts. Officials understood that if members were brought face-to-face with the full extent of their indebtedness, the shock might undermine their willingness to continue farming, or even cause them to leave the moshav. Officials also feared that members might react to the bitter revelation by ousting them from office.

The distribution of deficits and the regulation of members' private consumption were interrelated issues: The blurring of the true picture of members' debts through the accumulation of deficits in the cooperative's accounts was a means of self-deception and deception of others, instrumental in the rationalization of policies that, in general, allowed members to consume according to their apparent current income, without reference to their total real debt.

The following section details the overall effects of the aid process in one of the Biq'at Hayarden moshavim and the degree of success achieved in realizing the twin goals of distribution of deficits and regulation of private consumption.

## KFAR ZIMRI

Kfar Zimri was founded in 1975 by a settlement nucleus of sixteen young couples in their early and mid-twenties, mostly native Israelis of mixed ethnic background. The population did not grow very rapidly, and in the peak years during the period up to September 1984, there were thirty-five to forty families in the village. The farm unit provided to the settlers was based primarily on vegetables (50% to 55% of farm sales), with table grapes (33%) and flowers (10%) as secondary branches. Kfar Zimri is affiliated with the Jerusalem Corridor regional purchasing cooperative and is a member of the Labor-oriented Moshav Movement (Tnu'at Hamoshavim), the largest association of moshav-type villages in the country.

Both members and outsiders saw Kfar Zimri as a bad moshav: Its members were unable to get along with one another, and the settlement as a whole was unwilling to maintain certain basic norms of the settlement sector. The sources of the moshav's unsavory reputation are to be found in the moshbutz period; in keeping with the procedures then in force, the settlement operated during its first two years as a commune (kibbutz: thus, moshav functioning as a kibbutz equals moshbutz), with joint farm operations based on the members' labor and a uniform family living allowance. This arrangement was designed to help the settlers to gel as a community and also to provide them with practical experience in farming and cooperative management. Only in the third year were individual farm units allocated to the members' families. After receiving their own farms, the members began working on their own. It was expected that each family would consume at a level determined in keeping with its own private farm and nonfarm income.

The moshbutz period in Kfar Zimri proved to be a social disaster, and conflicts originating in that period continued to disrupt the social life and cooperative functioning of the moshav through the years. Factional conflict broke out when the Jerusalem Region of the Settlement Division tried to raise the level of the settlement nucleus founding the new moshav. According to the plan, Kfar Zimri was to be established by sons from new immigrant moshavim in the Jerusalem Corridor. However, the group based on this population failed to satisfy the region, and so it increased the village's population by bringing in additional settlers, in particular a number of kibbutz leavers. This resulted in the creation of two factions: settlers from moshav background, together with some ex-urbanites of lower educational levels, versus ex-kibbutz members and former city residents with higher educational levels. The latter group was referred to as the *kibbutznikim* or the "college grads," while the former was termed the "urbanites" or the "blacks," a reference to the oriental ethnic origin of the settlers from the Jerusalem Corridor moshavim.

The Settlement Division could not act directly and openly to remove from the moshav those settlers it considered less suitable. It also could not, or would not, intervene in the daily life of the moshav to advance the positions of the kibbutznik faction. Rather than improve the situation, the Division's intervention served only to create a conflict between two distinct, numerically balanced factions, neither of which could expel or overcome its rival.

The kibbutznikim, who had wished to realize a high level of cooperation, were forced to reconcile themselves to the maintenance of the lowest common denominator of cooperative practice in order to prevent open and active conflict with the opposing faction. The moshav acted to maintain organized marketing of members' produce, but, otherwise, the cooperative kept to a minimum its attempts to influence members' economic behavior. Members of both factions were embittered, felt alienated from the community, and withdrew from the public arena to cultivate their own private concerns. Without member involvement and support, the moshav cooperative was too weak to exercise effective control over member use of credit for private consumption. The cooperative was unable to supervise members' farm activities in such a way as to improve profitability. The social divisions which began in the first days of the settlement contributed, through the weakening of the cooperative, to the economic failure of the moshav.

The economic results of the early period just described were catastrophic. The audited accounts of the first five years of the village show that during the period 1975/76–1979/80, the excess of expenditure over income totaled roughly $1,140,000, or excess expenditure per family per year of about $9,400 (September 1984 value of the dollar). In terms of excess of expenditure over income, Kfar Zimri ranked second out of four moshavim that were studied in detail (Sherman, Sternberg, Eger, and Cohen 1987). During the period studied, there were only twelve moshavim all told in the Biq'at Hayarden.

The factor that allowed the members to withdraw from the public arena, despite the economic failure of the cooperative and most of the individual members, was the actual and expected flow of special financial assistance provided by the Settlement Division. This assistance allowed the moshav to function, despite the settlers' inability to create an effective cooperative structure capable of managing joint affairs and disciplining the individual members. Table 2.1 shows the timing and magnitude of the provision of funds to Kfar Zimri.

**Table 2.1**
**Special Financial Assistance, 1979–1984**

| Program | Sum | Date |
|---|---|---|
| Debt conversion "A" | $85,000 | 12-79 – 1-80 |
| Debt conversion "B" | $81,000 | 6-80 |
| Subsidized loans | $811,200 | 9-12-82 – 6-7-83 |
| Completion of payment of various settlement division obligations to Kfar Zimri* | $99,000 | 3-83 |
| Participation of the settlement division in special individual debt-liquidation program | $42,000 | 3-84 |
| Coverage of debts of settlers who left Kfar Zimri | $735,800 | 4-82 – 3-9-84 |
| TOTAL | $1,855,900 | |

*Coverage of "moshbutz" deficit; coverage of the deficit of the first year of operation on the basis of individual family farm units; subsidization of moshav cooperative operating expenses.

Though its assistance kept Kfar Zimri from collapse, problems arose in the realization of the Settlement Division's goal of reforming the management of the moshav through a combination of incentives and oversight. Indeed, a review of events in Kfar Zimri shows that despite intensified supervision, receipt of financial aid led to distortions in the handling of both cooperative deficits and private consumption, precisely the key topics singled out by the Settlement Division for improved management and stronger discipline.

The beginnings of the process of financial assistance to the settlements in the Biq'at in 1979 to 1980 found Kfar Zimri in its third year as a regular moshav and enjoying a very successful agricultural season. After several years of violent factional conflict, the members had come to a tacit understanding that the only way to live together in peace was to limit cooperative intervention in the members' economic affairs to a minimal level. Furthermore, the members discovered that they shared a common taste for higher levels of private consumption. The kibbutznikim, though publicly committed to a more rigorous adherence to cooperative norms, went along with steps in this direction initiated by representatives of the blacks–urbanites group.

The moshav prevented the returns from a bumper year from being absorbed by its large debt to the regional purchasing cooperative by the simple expedient of marketing its product to private merchants who paid cash directly to the moshav. Had Kfar Zimri acted as expected by the Division and the Moshav Movement and marketed through the national produce marketing cooperative, the latter would have paid the moshav by crediting its account in the regional organization, thus denying the moshav effective control over the proceeds of the sales of members' produce. Kfar Zimri's policy of selling to private businessmen kept farm income in the hands of the moshav and undermined the Division's attempt to enforce a cash flow plan in cooperation with the regional organization.

Meanwhile, most members' personal accounts in the moshav showed considerable credit balances as a result of the failure to divide all of the cooperative's ongoing expenditures and its accumulated deficits among the members. The settlers found the temptation provided by their credit balances irresistible, and the general assembly decided on division of "profits" from the 1979/80 crop year. A sum, variously estimated at $176,000 or $219,000, was thus distributed to twenty-three members. Shortly after the decision to distribute "profits," the moshav committee decided to debit members with the 1977/78 season's deficits according to their nominal value. Given the interest on these deficits, the meaning of this decision was to debit the members with only 21 percent of the actual debt accumulated.

The underlying factor that made possible a series of events of this nature was the flow of special financial assistance to the moshavim. Members justified the failure to debit cooperative deficits in full and the decision to distribute "profits" at a moment when the cooperative's situation vis-à-vis outside creditors was extremely difficult by claiming that the bulk of the cooperative's debt was in fact the responsibility of the Settlement Division. The Settlement Division did indeed acknowledge its responsibility for a considerable percentage of the moshav's debt. It suited the members to assume that when the Division had finished providing the expected assistance, it would be discovered that the financial situation of the moshav was not so bad after all.

The staff of the Settlement Division was infuriated when it became aware of the fact that Kfar Zimri's members had handed out to themselves large sums in cash against fictitious credit balances. The "profit" distribution came at a time when the Division was giving high priority to special debt relief measures and to a parallel attempt to intensify oversight of moshav financial management. In a rare display of determination, the Division responded by imposing disciplinary measures, limiting or denying the provision of various types of financial assistance during a period of almost two years. The Division relented only when it received concrete guarantees that the reforms in cooperative management that it demanded would be implemented; specifically, financial assistance was resumed only after the general assembly of the moshav had agreed to accept the installation of the Treasurer of the Moshav Movement as a member of Kfar Zimri's 1982/83 management committee.

The new management committee distributed the 1978/79 deficits according to their real value, a distressingly large sum that rocked the moshav and led a number of families to decide to abandon the village by the end of the year. Meanwhile, however, the flow of financial assistance from the Division began once again; indeed, the extent of the assistance provided to the moshav during the course of 1982/83 suggests that an attempt was made to compensate the moshav for the difficulties created by the imposed sanctions. In addition, the influx of aid together with another good agricultural year made it possible for the committee, led by the movement treasurer, to adopt a liberal consumption policy. Consumption was allowed to reach levels corresponding to current income rather than being held to a more modest level in keeping with the members' accumulated debts.

The social disruption created by the distribution of deficits in 1982/83 made it necessary for the Registrar of Cooperative Societies to appoint a management committee of outsiders for Kfar Zimri for the 1983/84 season. The committee consisted in effect of two members: one a senior representative of the Moshav Movement and the second a member of the headquarters' staff of the Settlement Division's Jerusalem region. The appointed committee undertook two major projects: It devised and imposed a debt-repayment program for those members most deeply in debt, and it continued the process of debiting members with cooperative deficits. The deficits presented to the committee by the moshav's salaried, nonresident treasurer and bookkeeper did not, however, include the moshav's interest account. The decision of the treasurer and the bookkeeper to leave interest out of the list of deficit figures was based on their belief that surpluses had accumulated in the interest account of later years and on their expectation that debt relief funds to be received from the Settlement Division could be applied to balancing the interest account. In other words, even when the management of moshav affairs was placed in the hands of representatives of the Settlement Division and other salaried professionals, the desire to avoid shocking the members and lack of clarity regarding the extent of the Settlement Division's financial commitments acted as obstacles in the way of full division of moshav debts to members' personal accounts.

The research findings presented show that Kfar Zimri was highly susceptible to the negative influences of the financial support provided by public institutions. Financial aid prevented the moshav's collapse but at the same time lured the members, individually and collectively, into acting irresponsibly. Neither individual ideological commitment nor social solidarity and community pressure were sufficient to guarantee that the members, in the management of their own private economic affairs or in the stands they took as members of the moshav cooperative, would overcome the temptation to seek to ease the burden of economic difficulties by passing them on to the care of national institutions.

Indeed, the case of Kfar Zimri shows that even when the Settlement Division was aroused, applied sanctions, and eventually became directly involved in the management of moshav affairs, the Division's dependence on the settlers was so strong that the basic problems of moshav financial management recurred: Consumption remained too high given indebtedness and members could not be forced to confront the full extent of their financial entanglement through a full division of deficits to private accounts.

Events in Kfar Zimri illustrate trends typical of the entire moshav sector in Biq'at Hayarden and in the country as a whole. The financial management of pre-1967 moshavim has not been explored in the same manner as that of the moshavim in the Biq'at Hayarden. However, the findings of a study of the moshav regional purchasing cooperatives indicate that pre-1967 and even veteran (preindependence) moshavim tended to waste resources to the extent that they felt that public institutions could be relied upon to rescue them from financial distress (Schwartz, Marvid, and Arian 1988). An economic study explains in similar terms the crisis of indebtedness that today encompasses the entire settlement cooperative sector, kibbutzim and moshavim alike (Zussman, Kislev, and Lerman 1989).

It should also be noted that there are exceptions to the patterns of behavior described in the village history presented previously. An important minority of moshavim are not affected by public financial assistance in the manner that characterized Kfar Zimri. Many of the moshavim founded before independence and some of the postindependence settlements demonstrate significantly higher levels of managerial competence and financial morality than those of Kfar Zimri and like settlements. In Biq'at Hayarden, the two most senior of the region's twelve moshavim did indeed function in keeping with basic managerial and moshav civic norms of financial responsibility.

Nonetheless, as the numbers indicate, these examples of successful moshav functioning are few, and the two positive cases exhibit important background characteristics differentiating them from the majority: both were founded by strong settlement nuclei whose members shared a certain ideological commitment and the experience of joint activity prior to their actual settlement in the Biq'at Hayarden. The two strong moshavim, established soon after the Six Day War, thus reflect in their origins some of the normative characteristics of pioneering rural settlement as institutionalized in the preindependence period. In the moshavim established later on, these characteristics tended to be either completely lacking or considerably weaker, as in the case described earlier.

## SUMMARY AND CONCLUSION

The history of Kfar Zimri illustrates the manner in which the continuing flow of public funds to rural settlements corrupts the basic cooperative institutions of the average moshav. With members committed to cooperative values and to their shared community only to a limited extent, the flow of outside funding has served to empty the moshav concept of mutual responsibility of its practical meaning. The expectation of emergency aid from public agencies weakens the pressure felt by members and management to conduct the cooperative's affairs in an efficient, orderly, and disciplined manner. The expectation of assistance undermines the ability of cooperative leadership and community social pressures to impose financial discipline on the individual members. Since the settlers' financial obligations are likely to be covered by an influx of public funds, one member is not likely to see in another's debt a direct financial burden. Demands directed to the heavy debtors are likely to be interpreted as unreasonable meanness and lack of sympathy for those who face the same difficulties and hardships shared by all.

Moreover, the attempt to impose discipline on members of the community by demanding greater efforts in production, restriction of consumption, and repayment of debts inevitably arouses social tensions and may even generate out-and-out conflict in the community. All those involved must necessarily ask themselves if it is worthwhile to pay the price of social disruption in order to maintain ideals of financial responsibility when the transfer of debts to the general public stands as a reasonably certain alternative solution. Given the lack of ideological fervor and social solidarity prevalent in today's moshavim, the answer is typically that the price is not justified: Members prefer to direct collective energies to obtaining external assistance rather than undertaking a direct, independent confrontation of individual and joint economic difficulties.

The situation in which Kfar Zimri found itself entangled was a reflection of the general crisis of the moshav sector and of the rural-agricultural sector as a whole. It seems almost certain that the coming years will see significant changes in the extent and terms of public financial assistance to the moshavim. The nature of such assistance deeply affects the functioning of the moshav, and so it may be assumed that the moshavim themselves will undergo important changes as communities and organizations as a result of changes in the public financial backing available to them.

Phenomena similar to those discussed previously have been noted in research on cooperatives in the developing countries. The clearest examples are to be found in the use of cooperatives as an organizational framework for the distribution of credit. In the absence of formal land titles, it has been suggested that cooperatives could distribute credit, with members' joint responsibility for cooperative debts guaranteeing loans, instead of mortgaged property. Such arguments were used, for example, in early postindependence Uganda in order to justify the rapid expansion of credit provided through cooperative societies. However, the concomitant political interest in gaining support through the issue of cooperative loans made the loans unconditional. The farmers sensed that this was the case, and thus the abundant flow of funds was accompanied by high nonrepayment rates. Ugandan cooperators, like their Israeli counterparts, could see no point in being harsh with one another as long as the authorities were being forgiving or weak. Only after the downgrading of cooperatives as a means of mobilizing political support and the severe restriction of the funds available did the loan recovery efforts of the Cooperative Department begin to bear fruit (Young, Sherman, and Rose 1981).

It is possible to generalize and state that when cooperatives are assigned high priority by government for political or development purposes they are likely to consider themselves immune to market discipline and to act accordingly. Policymakers interested in maintaining economic efficiency must find a way to post clear limits to the extent of public support for cooperatives.

## NOTES

This chapter was first published in 1991 as an article in *Human Organization* 50(2): 163–172. Reprinted with the permission of the Society for Applied Anthropology and from the journal, *Human Organization*. The authors wish to thank Levia Applebaum, Moshe Lissak, Yitzhak Taub, and the reviewers appointed by *Human Organization* for their valuable comments on earlier versions of this chapter. They also wish to acknowledge a special debt of gratitude to friend and mentor, Johnnie Sternberg, who shared his deep understanding of moshav accounting and moshav psychology.

1. Because of the political controversy surrounding settlement in the territories, the activities of the Settlement Department of the Jewish Agency, in which Zionist and non-Zionist Jewish organizations cooperate, are limited to the pre-1967 territory of the state. The Settlement Division of the World Zionist Organization carried out settlement activities in the territories occupied during the Six Day War. The budget of the Settlement Department is based on the contributions of Jews overseas, whereas the Settlement Division is funded as part of the budget of the State of Israel.

At the middle and upper levels, the staff of the Department and the Division overlap; that is, the same individual fills the same role in each of the juridically distinct agencies. In terms of organizational outlook and operational methods, the Department and the Division are to all intents and purposes the same organization.

In the text, the term *department* is used in general statements and in the historical review. The term *division* is used in the case study, since the moshav in question is located in the occupied territories.

## REFERENCES


Abarbanel, J. S. 1974. *The Cooperative Farmer and the Welfare State.* Manchester: Manchester University Press.

Baldwin, E. 1972. *Differentiation and Cooperation in an Israeli Veteran Moshav.* Manchester: Manchester University Press.

Ettinger, A. 1945. *With Hebrew Farmers in Our Land.* Tel Aviv: Am Oved.

Gvati, H. 1981. *A Century of Settlement: The History of Jewish Agricultural Settlement in Eretz Yisrael.* Tel Aviv: HaKibbutz HaMe'uhad Publishing House.

Gyllstrom, B., and F. Rundquist, eds. 1989. *State Cooperatives and Rural Change.* Lund: Lund University Press.

Harris, W. W. 1980. *Taking Root: Israeli Settlement in the West Bank, The Golan and Gaza-Sinai, 1967–1980.* Chichester: Research Studies Press.

Haruvi, N., and Y. Kislev. 1984. "Cooperation in the Moshav." *Journal of Comparative Economics* 8: 54–73.

Hyden, G. 1984. *Efficiency versus Distribution in East African Cooperatives.* Nairobi: East African Literature Bureau.

Kushner, G. 1973. *Immigrants from India in Israel.* Tucson: University of Arizona Press.

Mundlak, Y. 1964. *An Economic Analysis of Established Family Farms in Israel.* Jerusalem: Falk Center for Economic Research in Israel.

Nevo, N. 1978. "Social Change in a Veteran Moshav Ovdim." Ph.D. diss. Tel Aviv: Tel Aviv University.

Preparatory Commissions. 1929. *Key for the Settlement of Various Zones in Palestine.* Tel Aviv: Palestine Zionist Executive.

Public Committee for Settlement Affairs. 1966. *Report.* Jerusalem: PCSA.

Rokach, A. 1986. *Farm Types and the Settlement Key in Agricultural Settlement in the Land of Israel.* Rehovot: Settlement Studies Center.

Sadan, E., and D. Weintraub. 1980. "Ethnicity, Nativity and Economic Performance of Cooperative Smallholding Farms in Israel." *Economic Development and Cultural Change* 28: 487–507.

Schwartz, M., R. Bar-El, R. Finkel, and A. Nesher. 1987. "Moshav-Based Industry in Israel." In *Rural Industrialization in Israel,* R. Bar-El, ed., pp. 57–104. Boulder, Colo.: Westview.

Schwartz, M., A. Marvid, and O. Arian. 1988. *The Organizations: Three Moshav Purchasing Organizations Before and After the Crisis.* Rehovot: Settlement Studies Center.

Shapira, A. 1973. *Futile Struggle.* Tel Aviv: HaKibbutz HaMe'uhad Publishing House.

Sherman, N. 1979. "The American and Israeli Models of Agricultural Development: A Comparative Analysis." In *The Role of US Agriculture in Foreign Policy,* R. M. Fraenkel, D. Hadwiger, and W. Browne, eds., pp. 181–220. New York: Praeger.

———. 1980. "The Agricultural Sector and the 1977 Knesset Elections." In *The Elections in Israel 1977,* A. Arian, ed., pp. 149–170. Jerusalem: Jerusalem Academic Press.

———. 1982. "From Government to Opposition: The Rural Settlement Movements of the Israel Labor Party in the Wake of the Election of 1977." *International Journal of Middle East Studies* 14: 53–69.

———. 1993. "The Rural Settlement Cooperative Sector: From Ideological Symbol to Interest Group." In *Local Communities and the Israeli Polity: Conflicts of Values and Interests,* E. Ben-Zadok, ed., pp. 211–239. Albany: State University of New York Press.

Sherman, N., and M. Schwartz. 1987. Centralization versus Local Autonomy: Stalemate in the Settlement of a Politically Sensitive Region. Conference paper, XXVII European Congress of the Regional Science Association, Athens.

Sherman, N., M. Schwartz, and J. Sternberg. 1984. "The Guided Settlement Method in the Jordan Valley." In *Settlement and Regional Development,* A. Rozenmann, ed., pp. 291–304. Jerusalem: Keter.

Sherman, N., J. Sternberg, M. Schwartz, and R. Perlman. 1984. *Settlers and Settlement Agencies.* Rehovot: Settlement Studies Center.

Sherman, N., J. Sternberg, M. Eger, and T. Cohen. 1987. *Special Financial Assistance to the Biq'at Hayarden Moshavim.* Rehovot: Settlement Studies Center.

Shokeid, M. 1971. *The Dual Heritage Immigrants from the Atlas Mountains in an Israeli Village.* Manchester: Manchester University Press.

Weingrod, A. 1962. "Administered Communities: Some Characteristics of New Immigrant Villages in Israel." *Economic Development and Cultural Change* 11: 69–84.

———. 1966. *Reluctant Pioneers.* Ithaca: Cornell University Press.

Weintraub, D. 1971. *Immigration and Social Change.* Jerusalem: Israel Universities Press.

Weintraub, D., M. Lissak, and Y. Atzmon. 1969. *Moshava, Kibbutz and Moshav.* Ithaca: Cornell University Press.

Willner, D. 1969. *Nation-Building and Community in Israel.* Princeton: Princeton University Press.

Young, M. C., N. Sherman, and T. H. Rose. 1981. *Cooperatives and Development: Agricultural Politics in Ghana and Uganda.* Madison: University of Wisconsin Press.

Zussman, P., Y. Kislev, and T. Lerman. 1989. "Credit for the Moshavim—The Experience and Its Moral." *Economic Quarterly* 40: 114–131.

# Chapter 3

# The Settlement Department Unsettled

*Shlomo Fox*

*Shlomo Fox documents changes in the relationship between the Settlement Department of the Jewish Agency (a quasi-governmental institution) and the moshavim by describing three cases at three successive time periods. The progression seems to go from paternalism to a kind of benevolent disinterest after an unsuccessful excursion into experimentalism with industry. Has the very success of the Settlement Department's earlier ventures rendered it obsolete in a now well-developed countryside? Or has the decline in the importance of agriculture to the national economy led to its irrelevance? Or, yet again, has a paternalistic intervention became outmoded in Israel's modern economy and political climate?*

The Settlement Department of the Jewish Agency has been the principal bureaucratic instrument for rural development in Israel for at least the last fifty years. At the peak of its activity in the 1950s, it engaged more than 2,000 employees in the process of absorbing thousands of immigrants into the rural sector within the framework of new moshavim. By the end of 1992, the Settlement Department[1] became extinct.

It is the intention of this chapter to argue that the demise of the Department parallels a gradual and radical change in its relationship to the rural sector, roughly a change from commitment to acquiescence, from initiating to reacting.

In general terms, the change pivots around a number of factors. These include the national transfer of political power and ideological emphasis at the end of the 1970s; the transfer of the Department's attention, effort, and resources from moshavim and kibbutzim to community settlements (*yishuvim kehilatiim*); a transfer of priorities in favor of the territories of Judea and Samaria; and, perhaps most important, a transfer of interest from agriculture to industry.

The particular arenas and periods from which these conclusions are drawn are respectively a moshav in the Lachish region in the early 1970s, a moshav in the Jerusalem Corridor from the mid-1970s to the early 1980s, and community and collective settlements in the Hebron area through the 1980s.

The critical responsibility that the Settlement Department took upon itself with respect to those new moshavim built for and from the newly arrived populations of the 1950s was economic. This responsibility expressed itself as a commitment to the economic viability of the elemental family farm, the elemental cooperative (moshav) and, inevitably, to the regional and national institutions of the moshav sector. It expressed itself most clearly and practically in the Department's design of regional farm archetypes, each adapted to resource and climatic conditions, such as industrial crops in Lachish or poultry in the Jerusalem Corridor, and was replicated systematically in the constituent family farms of the moshavim. The archetypal farm was planned to generate income at a level no less than that earned by the urban industrial worker.

Moshav Noam can serve as an example. The moshav was founded in 1955 and is situated a few miles south of Kiryat Gat. It was part of the Lachish regional development program. The moshav was initially planned for eighty family farms, each with thirty dunam of irrigated land, 18,000 cubic meters of water, and a standard parcel of implements and equipment. Each was expected to grow industrial crops, such as sugar beets and vegetables. The cropping pattern was intended to generate a level of net disposable income comparable to urban industrial levels.

With the decline of the profitability of sugar beets in the late 1960s and its eventual demise as a domestic industrial crop, livestock was introduced into Noam's farm development program: dairy farming and turkey production for the larger, demographically mature families and young couples, respectively.

The Settlement Department provided almost all the finance for the required livestock, buildings, and equipment. It planned the location of the cowsheds and turkey pens, purchased the livestock, and provided the farmers technical expertise through local instructors. The objective of establishing such turnkey farms was to develop a rural economy and community strong enough to withstand the centripetal attraction of urban life, particularly in the main metropolitan areas.

If the case of Noam in the early 1970s exemplifies how the Settlement Department practiced its commitment to the viable rural economy, the case of moshavim in the Jerusalem Corridor through the mid-1970s and early 1980s illustrates how the commitment was preserved normatively, while in practice, the design of the family farm implied significant commuter employment.

Messilat Zion is a moshav just off the main highway between Jerusalem and Tel Aviv and roughly twenty-five minutes from each. It was established in the 1950s by immigrants from Cochin, India and, together with some dozen moshavim, occupied the narrow hilly corridor which at that time joined Jerusalem to the coastal plain.

The farm model for these moshavim was comprised of poultry and orchards—specifically in the case of Messilat Zion, 1,500 laying hens and ten dunams of fruit trees. By the beginning of the 1970s, poultry farms in the Jerusalem Corridor were small-scale, fragmented, poorly equipped, nutritionally inefficient, environmentally offensive, and of difficult accessibility. In the early 1970s, the Settlement Department began to implement a large-scale expansion and modernization of poultry farming in the Jerusalem Corridor within the framework of its commitment to the farm model. At Messilat Zion, more than one million dollars was invested over five years to relocate the poultry pens away from residential areas and to provide new buildings and equipment.

The necessity of the project was argued in terms of minimum veterinary and public health requirements and the creation of efficient input delivery systems to ensure economic survival.

Having been closely involved in this and other like projects, it is possible to recall the uneasy discrepancy between the massive input of financial resources, professional manpower, and plain emotional commitment that we invested and the realization that the central premises of the farm model would not be attained—that no more than one-third of target income would accrue from a family egg farm, nor no more employment than would occupy someone part-time, probably the farmer's wife.

The logic of our objectives demanded more, and the shortfall was addressed in a number of ways. While the outcome in terms of the family farm structure that emerged was remote from the Department's normative model, it is central to the argument that the model provided legitimacy to the measures that were taken by which the shortfall was addressed.

First, while the Department built pens for 1,500 layers for each family, land was leveled to enable farms to expand eventually to 5,000 layers, which, if ever achieved, would realize handsomely our income and employment objectives.

Second, a number of elements in the family farm model were pooled to create large projects, centrally managed by the cooperative, with permanent staff and temporary seasonal labor, not all of whom were from the village. Such was the case with orchards. The farmer's model stipulated ten dunams of orchards for each family. In practice, village orchards—apricots, avocados, almonds, and peaches—were developed by the Department as single units where the involvement of most families was slight except for perhaps a few days of fruit picking a year and the annual allocation of any net profits or loss.

The concentration of family orchards in this way was doubtlessly the most efficient way to manage them. Likewise, the village centrally managed cereal and cotton production. Even the rearing of laying hens from chicks was removed from the family farms, where it had been until the poultry redevelopment project, and concentrated in large, modern, climate-controlled growing houses under central village management.

Third, as part of a national and regional development thrust in the early 1980s, flower production for export, mainly carnations and roses, was encouraged by the Department. While the income and employment projections promised the achievement of the Department's goals, if not from one dunam than from two, it was recognized that flower farming would not be a blanket solution for all farms at Messilat Zion in the way that egg production was. A process of selection was implemented with candidates undergoing a series of aptitude tests. Perhaps the most significant departure from previous practice was that farmers initiated their own candidature and thus for the first time were taking responsibility for the development of their farm businesses.

While Settlement Department planners may have agonized over farm design and resource input at Messilat Zion, no such reservations applied to the villagers themselves. On the contrary, the development path chosen by the Settlement Department for Messilat Zion allowed most of them to operate a small poultry farm, providing a useful supplement to the family income, requiring no more than a few hours each day of light work distributing food and collecting eggs, and allowing employment off the farm in the orchards or brood houses, in nearby agro-industrial enterprises, or in Jerusalem and Tel Aviv. A large proportion of farmers earned a significant part of their income from such outside employment, mostly in town.

If the moshavim of the Jerusalem Corridor were pragmatic communities enmeshed in a system of bureaucratic objectives and establishment ideology, something of the opposite applied to the community settlements that were initiated in the territories of Judea and Samaria after 1977. These communities embodied in their very location, and frequently in their local culture, their own strongly held political and religious values. The Settlement Department, for its part, was reduced to providing professional planning services and development financing.

Adora is a community settlement[2] established in the mid-1980s several miles west of Hebron on the main road going down to the coast. Its closest neighbor, a collective moshav called Telem, had been established a few years before on the same stretch of road. It is instructive to compare the way the Settlement Department dealt with these two villages in order to demonstrate the operation of bureaucratic commitment.

First, though, consider their similar "genetic" data:

1. Same location and proximate founding date
2. Same political patrons: Herut-Betar
3. Similar population: secular, urban, locally raised

The critical difference between the two is in their economic structure. Adora is a composite of economically autonomous households, each free to try whatever job and business opportunities, wherever, and unambiguously responsible for the outcome. Telem is a single centralized economic entity presenting little employment flexibility to its members but taking responsibility for their livelihood.

The distinction between the two villages in terms of Settlement Department involvement is clearly defined. As for Adora, the Department deals mainly with the processing of applicants for membership, with the financing of certain infrastructural investment; it oversees village management and finances part of its current administrative budget. The Department does not initiate economic projects in the village, although it is interested in business proposals from villagers and is prepared to finance for them the construction of industrial buildings and investment in industrial equipment. For several years after its inception, little business activity occurred in Adora, and most people commuted to Kiryat Gat, Kiryat Arba, Jerusalem, and even further. Over the last three years, a number of small businesses have been established, mainly in the field of metal and mechanics, and 900 cubic meters of industrial buildings have been erected. The village has currently around sixty employable members, of whom six are employed by local businesses.

This situation of negligible local employment has not been a matter of serious concern for the Settlement Department. At Telem, on the other hand, from its inception, the Department has made a serious sustained effort to actively develop local employment and income. By 1988, the village was operating a small vineyard, a chicken farm, and struggling to maintain a factory producing extruded rubber products. This last in particular epitomizes the unrelenting commitment of the Settlement Department to the creation of profitable employment for the village.

For the Department, the agricultural projects were a relatively simple operation, given its long accumulated experience. Agriculture, however, had little to offer, with no special skills possessed by the villagers and with the local and national constraints of climate, natural resources, and markets.

The decision to purchase the rubber extrusion plant—a going concern—was taken by the Settlement Department for Telem with the assent of the Herut-Betar organization. The Department financed the purchase totally, erected a building (700 $m^2$) to house the plant, engaged the previous owners to manage it for a transitional period, sent village members to technical and managerial training courses, paid outside experts for technological and marketing advice, sent people abroad to examine new technologies and, indeed, encouraged and partly financed the introduction of a new production line. By the time the business had failed in 1990, the Department investment had exceeded $800,000.

The contrast between these two villages in terms of Department commitment is clear and unambiguous. Telem and Adora, so alike in many respects, represent the old traditional, committed, development posture of the Department and the new passive posture, respectively.

It is worth noting that once the Settlement Department reached the conclusion that Telem was not viable socially and economically, it moved to transform it from a collective moshav to a community settlement. This process having been completed, the previous commitment to the economic development of the village evaporated.

It is sometimes maintained that the community settlement is descended from the moshav: both comprise families as independent economic units; both ascribe to an ideology of mutual help and self-labor; both facilitate finance for the economic development of their members; both develop and manage jointly owned productive assets (e.g., orchards, industrial buildings); and both constitute a municipal entity.

These points have some validity, but it is an error to portray the community settlement as a normative development of the moshav.

The causal vector from moshav to community settlement lies in the plane of pragmatism. The practical break with the principle of comprehensive commitment to the family economy as demonstrated at Messilat Zion becomes in the community settlement a normative change, a new (parallel) bureaucratic ideology, which gradually becomes dominant as the emphasis on rural development shifts to the community settlement in the hills of Galilee, Judea, and Samaria, where institutionally initiated farm models give way to private and relatively unique industrial enterprises. Any vestiges of the old ideology of total commitment were swept away by the avalanche of debt which struck the rural sector in the late 1980s and against which the Department interventions were irrelevant.

With this new passive posture, which had gradually infiltrated dealings with the moshavim and became legitimized in dealings with a more resourceful population in the community settlements, the Settlement Department lost its special multiplex relationship with the rural community, leaving it as just one more technical consultant/regional planner/manpower assessor among others in government and the private sector and exposed it as a willing but marginal source of soft finance in the area of nonagricultural business development.

## NOTES

1. The Settlement Department in this chapter includes both the Settlement Department of the Jewish Agency and the Settlement Division of the World Zionist Organization. The former is financed by world Jewry and does not operate beyond Israel's 1967 borders. The latter operates beyond the 1967 borders and is financed by the government.

2. The community settlement emerged as a distinct rural entity from the late 1970s, specifically after the change of government in 1977. Its roots are to be found partly in the rural service centers established in the Lachish region and elsewhere in the 1950s and partly in the Mitzpim in the Galil.

Above all, this type of settlement is associated with the territories of Judea and Samaria, where the predominant objective was to increase the Jewish population at a time and in places where agricultural development was rarely possible.

The community settlement is thus conceived as a commuter village, unconstrained in size by prospective agricultural resources and typically within tolerable commuter range of the city.

Chapter 4

# Socialism, the Moshav, and the Water Crisis

*Susan Lees*

*The success of the moshavim was, like that of all agricultural settlements in Israel, dependent on the provision of reliable, good quality, inexpensive water for irrigation. Conversely, Israel's water supply system was developed and organized primarily to meet the needs of its agricultural settlements. Here, Susan Lees looks at what lies behind contemporary changes in attitudes and policies governing the provision and allocation of water. As financial pressures mounted in the early 1980s, a series of drought years forced both rural and urban Israelis to consider whether to continue to allocate a generous share of its water to farming. While the argument has an economic component, politics and ideology were not far from the surface.*

Israel is never short of dramatic crises to occupy the mind of the public, and water crises are recurrent featured highlights on the dramatic scene, playing a strong third to war and economic catastrophe. In 1992, the water crisis was a matter of "too much," but more chronic is the type of crisis of the previous year, "too little." Water shortage, in turn, is not entirely unconnected with national security and the economic situation of the country. Now, in a period of political, economic, and ideological transition, the connections between these are being reexamined in light of the changing role of agriculture, and specifically moshavim and kibbutzim, in Israeli national life.

Given the kind of life that the founders of Israeli society envisioned for the new nation, development of and control over the meager water resources of the land was a central priority. For not only was it necessary to supply adequate water for urban residents to use for domestic purposes and industry, but more important, it was considered essential to provide plentiful water for the development of highly productive irrigated agriculture to the far corners of the land. And indeed, Israel's success in the development and management of its water supply had been exemplary, systematically studied by students from water-poor Third World countries all over the world.

In order to understand the Israeli achievements and current predicament in the matter of water supply, it is important to draw together many threads in the fabric of Israel's history. Since our story's main character is the moshav, we will, of course, focus more on some aspects than on others, but our main task will involve placing the moshav in the wider context of the society that gave it birth and may now be bringing about its demise.

## SOCIALISM AND EFFICIENCY

The argument can be made (and has been made) that socialism leads to inefficient use of resources. Inefficiency rises out of certain central tenets of socialism:

1. Socialism should promote cooperation, not competition; therefore, if individuals do not compete, they are not encouraged to make more efficient use out of what they have in order to succeed.
2. In socialist societies, the weak should be supported by the strong. Therefore, resources are expended inefficiently on poor users who waste them in unproductive consumption. Further, socialism abhors great differences in rewards. Since great disparities in incomes and lifestyles between individuals should be avoided, there is little incentive for anyone to work harder than others, as the rewards will be more or less equally distributed.
3. In socialist societies, everyone has a right to a job and the means to do it. This means that poor and inefficient workers are not "weeded out" and, again, waste resources. It further means that workers are not held accountable for their actions, as they do not have to pay the price of their inefficiency by losing their job or livelihood; hence, they have little motivation to do their jobs well (efficiently).

What all this comes down to, according to the argument, is that since, under socialism, the rewards of effort are weakly, at best, related to expenditure of effort, people are disinclined, through lack of motivation, to work hard to make the most of what they have. Instead, they regard what they have or are getting as an entitlement, a right, that they receive simply by virtue of being a member of society. As long as they continue to participate as a member by doing the minimum needed to qualify for membership, they will continue to receive a more or less equal share of what the society has to offer. This applies to basic resources such as land and water. A member will not lose entitlement to a share of land or water simply because it is not being used to greatest efficiency; furthermore, his or her standard of living will not be dramatically affected by better or worse use of the resource.

In Israel's case, the distribution of water for irrigation, an admittedly scarce, costly, precious resource, socialist principles prevailed. These are best illustrated in the moshavim. Nearly universal access to water for irrigation depended on membership in a cooperative community; entitlement to water went along with entitlement to land and other community membership benefits—indeed were inseparable from them. Water was rationed to communities on the basis of need as determined by the central management in the Ministry of Agriculture. This need was calculated in large part by use of a politically derived formula which took into account size, population, ecology, crops, and past history of use. While water allocation fluctuated from year to year according to availability and shifts in usage, it remained proportionately roughly as set in the 1960s (Figures 4.1, 4.2, and 4.3).

What is critical to the distribution of water is that the price that farmers paid for the use of water was not directly related to the cost of its production. Overall, the state subsidized about two-thirds of the cost of water production. This was unequally distributed between the north and the south—that is, a greater subsidy was paid for provision of water to the south so that farmers in the south would not bear a greater burden of cost to live and work there than farmers in the north, where water was more plentiful and cheaper. Thus, it has been argued that farmers in the south, where water is more scarce and more costly to produce, are not motivated to make more efficient use of water by the cost of production; hence they use it wastefully (Sadan and Pohoryles 1979; Galnoor 1980).

But though this reasoning applies conspicuously in the south, the overall subsidy has the same effect throughout the country as long as the cost of water production is not directly reflected in the

**Figure 4.1**
**Irrigated Area and Water Input (Millions of Cubic Meters)**

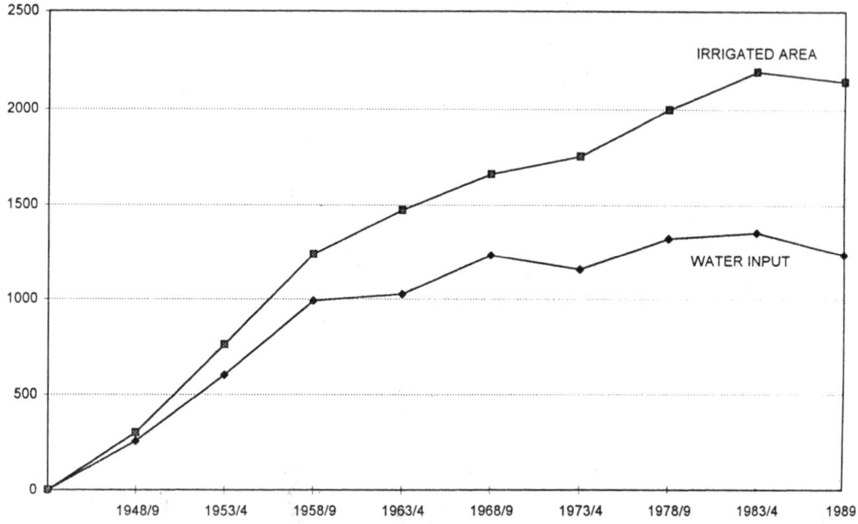


*Source: Statistical Abstracts of Israel* (1990), Table 13.1, pp. 388 and 393.

**Figure 4.2**
**Agricultural Water Input**

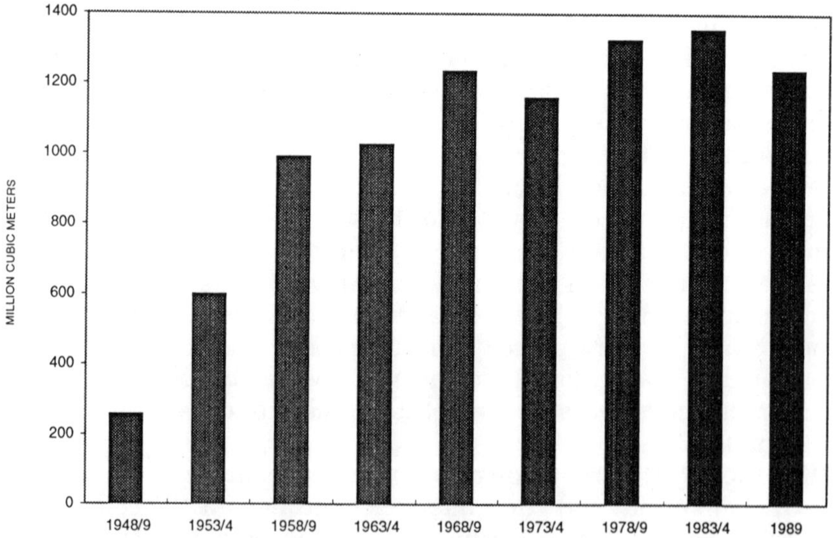


*Source: Statistical Abstracts of Israel* (1990), Table 13.1, p. 393.

**Figure 4.3**
**Total Water Consumption by Sector**

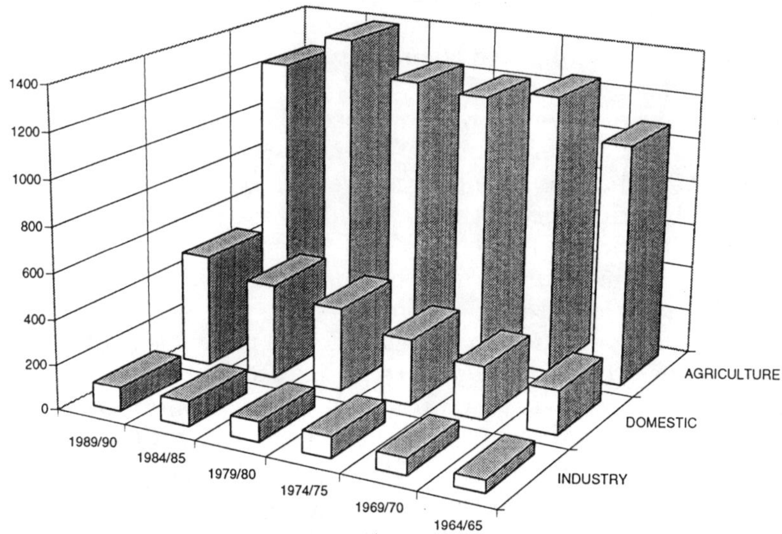


*Source: Statistical Abstracts of Israel* (1990), Table 15.6, p. 449.

price the farmer pays. A guarantee of equality (or at least equity) among farmers prohibits such a reflection, for geographical differences would always lead to differences in the cost of water production.

How much each community gets is in part determined by its size on the principle that every member is entitled to an equal share of the community's total allocation. Whether each individual actually needs an equal share is not an issue as long as equality in land and water allocation is a principle of community membership. Actual equal allocation of water among community members would, of course, lead to great inefficiencies in that, clearly, members differ in their agricultural needs and capacities through time with their family developmental cycle and with changes in their agricultural and other occupational activities. Cognizant of this fact, community members do adjust water allocations among themselves, as they do other agricultural quotas determined in bureaucratic manner by the central Planning Department of the Ministry of Agriculture.

Although the total quantity of water available to the agricultural sector as a whole should be regarded as limited in that there is a limit to how much can be safely withdrawn from the nation's aquifers in any one year, the principles of allocation lead farmers to treat the resource as if it were limited not by nature but by bureaucrats. Farmers do not compete with one another for the use of a limited resource by, say, bidding against one another according to how much they are willing and able to pay but, rather, consider themselves entitled to a share because they are members of farming communities. Water is not a marketable commodity; hence it is not regarded as a limited commodity by the users. I shall return to this point.

Some analysts have drawn a picture of Israeli water use along these lines: It is wasteful and inefficient because it is not controlled by market mechanisms that force users to be accountable on the basis of the profitability of their expenditure. Crises of water shortage draw attention to the defects of this system of water allocation because the entire country is threatened and pays the price in cutbacks in domestic water allocation.

However, anyone familiar with the actual practices of farmers in Israel and elsewhere in the world will have to concede an astonishing paradox: Israeli farmers are in fact among the world's most efficient users of water and have increased their efficiency of water use through time. Evidence of this is clear: While the allocation of water to the agricultural sector has remained steady or increased only very slightly, the quantity of land irrigated and the productivity per cubic meter of water applied has increased markedly (see Figures 4.1 and 4.2). In other words, despite all expectations arising from the previous description of socialist principles of allocation, Israeli farmers show evidence of using water efficiently, particularly in the small-holder sector (which received less water per farmer and per hectare and per community than did the kibbutz sector) (Figure 4.4). Now, why should this be the case?

Mitigating the anticipated negative effects of socialism on resource use, there were a variety of policies regarding water which produced, in fact, the reverse results. These policies promoted water-efficient practices only marginally through negative sanctions and very centrally through positive rewards (Laster 1980). The negative sanctions involved graduated pricing beyond the allocated quota: Once a community had exceeded its quota, the price of additional water rose very steeply (but it should be noted that until 1986, additional water was not systematically withheld). Because most communities had some farmers who did not use up their own shares of the allocation because they did not farm all their land, others could take their share at the subsidized within-quota price—hence, the pricing mechanism had only very limited effectiveness. However, it should be noted that communities rarely exceeded by very much their total allocations. In fact, until the 1980s, the total agricultural sector water use was less than the total allocation as often as it exceeded the allocation. This suggests that

**Figure 4.4**
**Water Consumption in Agricultural Communities**

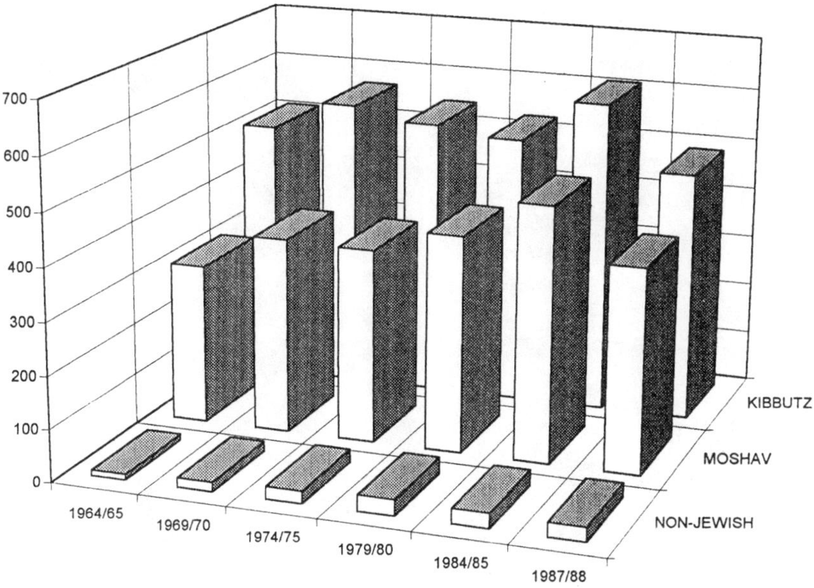


*Source: Statistical Abstracts of Israel* (1990), Table 15.6, p. 449.

even the limited effectiveness of the pricing mechanism was roughly adequate to guide the agricultural sector in the quantities of water that it could appropriately use.

Perhaps more effective and remarkable were the positive incentives devised to promote efficient water use. These were largely directed toward encouragement of the development of effective technological means of improving water use efficiency and dissemination of both information and means of implementing the new technology. Government policies included protection of infant local industries producing water-efficient technology, provision of low-cost loans to farmers adopting such technology on an experimental basis, and support for research in this area for education programs to disseminate and expand technological expertise among farmers. Israel quickly built up a worldwide and well-earned reputation for this expertise. The buildup of a cadre of trained experts in efficient irrigation techniques had its own payoff in that the profession had a momentum of its own, adding achievement to achievement and accelerating the pace of development (Figures 4.5 and 4.6).

**Figure 4.5**
**Irrigated Area as Percentage of Total Cultivated Area**

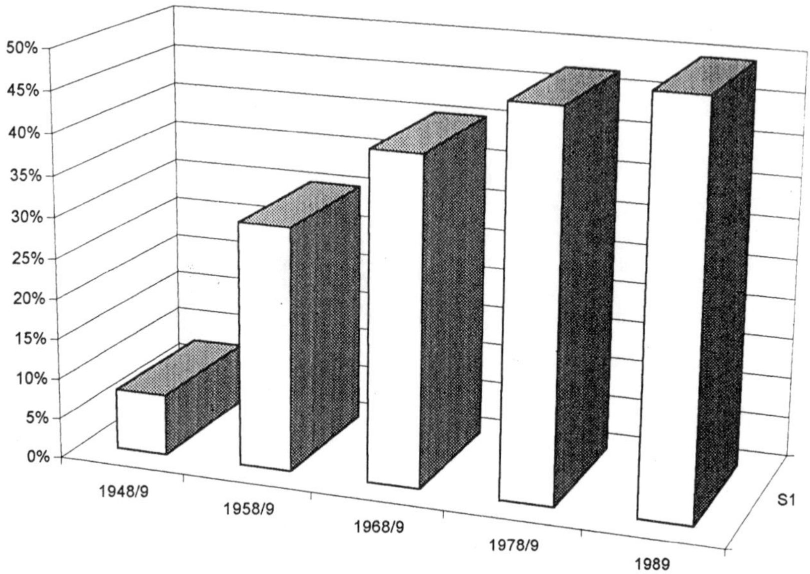


*Source: Statistical Abstracts of Israel* (1990), Table 13.1, p. 388.

But this approach to increasing the efficiency of water use came at a high cost. That is, technology does not come to the farmer for free, and keeping up with the most recent technological developments is costly to the farmer indeed. Even with government subsidies, farmers were obliged to go deeply into debt to keep their equipment up to date. In the 1970s and up to the early 1980s when inflation was high and money flowed rather freely, farmers invested heavily in farm equipment. When the austerity budgets of the mid-1980s were applied to bring inflation to a halt and curb indebtedness, farmers found themselves unable to make their productivity pay for their earlier investments, much less keep up the levels of investment of the past. Thus, commitment to high-tech irrigation, no matter how efficient the technology in terms of water use, contributed significantly to economic disaster in the 1980s.

The predicament of farmers, particularly moshavnikim, resembled that of family farmers in Europe and North America in that the market value of their crops could not keep pace with the debts arising from their commitment to new technology (Kramer 1987). Their problem was all the more acute because their farms

**Figure 4.6**
**Cultivated and Irrigated Areas**

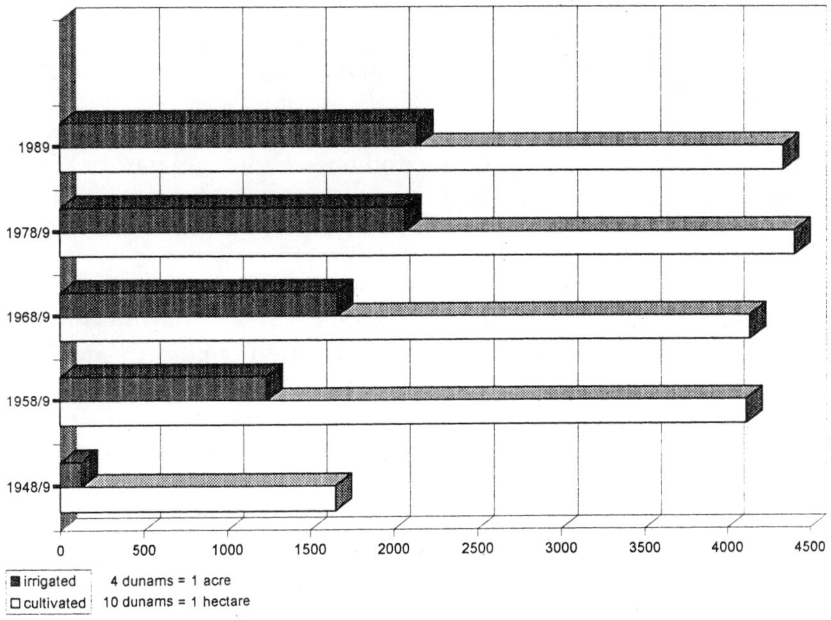


*Source: Statistical Abstracts of Israel* (1990), Table 13.1, p. 388.

were so minuscule. But while the European and North American family farmers experienced their indebtedness only as individuals, the moshavnikim, as members of a cooperative, drew their fellow members into debt with them so that even the successful farmers were liable for the debts of others. Being a good, clever, thrifty, honest, hard-working farmer was not enough.

If the economic problem was compounded by the socialist organization of farming, it did result in consequences that diverged sharply from those of Europe and North America. The proportions of farmers in the economy at large decreased everywhere, as did the numbers of individuals and families making their living by farming as a principal occupation. But in Europe and especially in North America, family farming gave way to huge, corporate agribusiness, which accelerated the demise of small-scale farming through competition. This competition did not rest on efficiency of productive techniques or use of land and water resources but on effectiveness of acquisition of credit, which large corporations can obviously do much more readily than small farmers (Kramer

1987). In Israel, acquisition of credit rested on the effectiveness of the regional purchasing cooperatives; when they failed in 1986, there seemed to be nowhere else to turn.

If the Western world was suffering from wave upon wave of agricultural crises in the 1980s, often blamed by the rural sector on the policies of rightist governments, the nations of Eastern Europe and the former Soviet Union were also having difficulties, these blamed on faulty central planning and on socialism in general. The situation for parts of the Third World was utterly catastrophic, particularly for most of Africa. While Israeli agricultural problems can be understood partly in terms of the efforts small-scale farmers were making to keep up technologically and financially in the context of scarce water resources, in fact, their situation did not differ very radically from that of small-scale farmers in much of the world (see, especially, Sexton 1990). This situation was the outcome of the contraction and marginalization of agriculture itself in the world economy. It had no particular relationship to efficiency or inefficiency; that is, the number of farmers running their own farms successfully was reduced worldwide (Figure 4.7), regardless of how efficient individuals were in their use of land, water, labor, or other factors of production.

## The Water Crisis

If we accept that Israeli farmers tended to be surprisingly efficient in their use of water, a second paradox in the management of water arises out of its socialist organization. That is, despite a very high level of efficiency in water use, the water resources themselves were being rapidly depleted. This is a crisis that goes far beyond the economic one in that it is harder to correct and has farther reaching consequences, even life-threatening, some would argue.

Shortly after the founding of the State of Israel, its Water Law was implemented, securing for the nation the right to control the allocation of all ground and surface water in the country. The purpose of developing Israel's water resources was agricultural development. This purpose is clearly articulated in Israel's Water Law, which not only locates control of the national resources in the Ministry of Agriculture but states that the use of water is to be devoted to national development and empowers the organ of the Ministry, the Water Commission, to take measures to achieve that goal. The constraints are two: Water is to be used efficiently and equitably. This is interpreted to mean that means are to be found to provide water equitably to those who are far from the sources and that users are to be encouraged to find ways of being more productive

**Figure 4.7**
**Extent of Agriculture and Agricultural Employment**

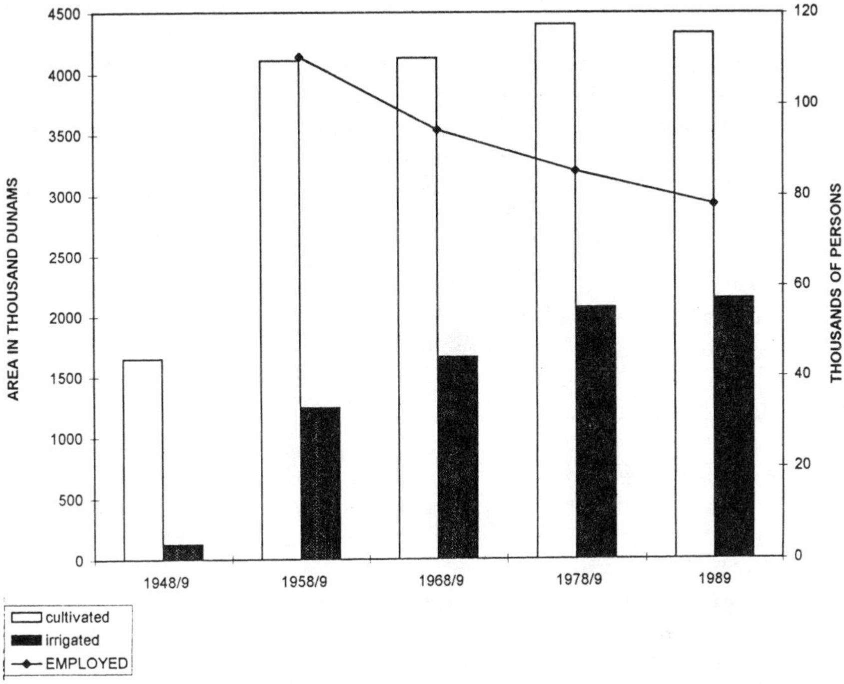


*Source: Statistical Abstracts of Israel* (1990), Table 13.1, p. 388.

and ways of avoiding waste. Two main devices were developed to implement these objectives: allocation and pricing.

Allocation depends in part on how much water is available. Estimating availability is not a simple matter, for it means calculating not only the storage capacity of underground reserves, flowing surface resources, and natural and artificial reservoirs but calculating the amount and rate of annual replenishment. In an arid climate with unpredictably fluctuating rainfall, this presents some uncertainties. Decision makers might borrow in some years, gambling on replenishment in subsequent years, hoping for overall stability in the long run. In fact, this is what contributed to the severity of the crisis in the 1980s, when six years out of seven were very low-rainfall years and provided too little water for both natural and artificial recharge of the reservoirs. A further complexity relates to the question of quality, namely, what degree of salinity is deemed acceptable? Hence, to what extent are saline sources to be included as part of the total water available?

The early planners of the water supply system estimated the total capacity of the system as they defined it, initially recommending provision of water at levels lower than the full potential because demand had not yet developed, then at increasingly higher levels to peak at a point that was substantially greater than the total amount available to the system in the long run. That is, it was decided initially to deplete certain resources on a one-time basis during the development period while the level of water use in agriculture was being increased. At the same time, it was also assumed that gradually, alternatives to the freshwater resources would be developed so that if agriculture were to sustain a high level of water demand, it would be met by replacements, especially from treated urban waste. The underground reserves, which eventually provided nearly all of the freshwater used, would be replenished after the initial depletions of expendable reserves by natural and artificial recharge on an annual basis; any water needed beyond that point would have to be added from other sources. The early planners estimated that the maximum level of exploitation of the aquifers could persist safely for thirty years, but just to be safe, they informed policy makers that they had only fifteen years in which to find and shift over to alternatives. In fact, the plan was put into effect but the development of alternatives and the shift-over was slower than circumstances and planning required, perhaps because of the dependence on treated waste water, which fell into the domain not of the Ministry of Agriculture but the local municipalities, Ministry of Health, and Ministry of the Interior. Even with treated sewage and the use of saline water, by the 1970s, Tahal was projecting that by the year 2000, agriculture would have to get along on a greatly reduced allocation.

National-level development of alternative resources has been slow (Issar and Nativ 1988), and lack of finances has been no small part of the problem. But lack of finances is the result of a political decision. Legislation to encourage urban centers to treat their own sewage has also been slow in coming, and farmers are charged high prices for treated sewage for irrigation, despite the fact that treated sewage has more limited use than freshwater. In other words, since the late 1970s, the government has shown little interest in encouraging changes in farming practices that will reduce the exploitation of freshwater resources and increase dependence on alternatives.

While the central government was proving less effective in keeping up with the plan for shifting agriculture away from dependence on the nation's freshwater resources, the Water Law makes provisions for strong incentives for local parties to develop their own resources. For example, the law provides that local developers of resources will not be penalized through reductions in their supplies should they provide new sources. In the north especially, regional water associations began taking the lead in developing local sources of water, including both seasonal runoff and treated sewage. The regional water association of Emek Yisrael is a good example. Tahal, the national water planning agency, and Mekorot, the national water supply agency, developed a large waste-water treatment system to convert Haifa's sewage for agricultural use in the valley. But Palgei Mayim, the regional association, built in the meantime some seventy-three reservoirs from local sources. The Haifa system was costly, time consuming, and technically and environmentally problematic. Palgei Mayim argued convincingly that building local reservoirs was more effective, cheaper, quicker, indeed, superior in every way in dealing with the water problems of the valley and finding solutions. Similar circumstances prevail in five or six other regions with strong water associations.

While the Water Law provides the potential for initiatives, and some incentives are there, with increasing threats to local water supplies, battles between local and national agencies, underfunding, and a lukewarm attitude by government have combined to slow the necessary push for development. This certainly reflects a shift in ideology, as neither the support of agriculture nor a sensitivity to the environmental necessity for sewage treatment are accorded priorities by the current regime. As long as these are regarded as national responsibilities, the lack of ideological support will contribute to a lag; private and regional initiatives will be inhibited by conflict with centralized jurisdictions.

At the same time, while there have been measures taken to reduce the share of water allocated to agriculture, these measures also have been halting and somewhat inconsistent. In 1986, the first serious national water emergency measures were enforced, with a 15-percent national reduction in water allocations strictly enforced throughout the country. Communities met to discuss the ways in which they would enforce the reduced allocations among themselves and the likely outcomes of future reductions on their agricultural strategies. It is difficult to know what would have happened had the drought continued. The winter at the end of 1986 and beginning of 1987 brought heavy rains, while the financial crisis persisted. Thus, the debt situation attracted far more attention than the water problem, obscured by the weather, which did not, of course, alleviate the problem, only its visibility.

The debt crisis, however, contributed in its own way to the alleviation of the water problem in that financial pressures led many to leave farming and others to reduce their farming dependence. The total water share of the farming sector was reduced. However, the vast influx of new immigrants from the Soviet Union exacerbated the problem because of sharply rising urban demands for water and because of severe pressures on the national economy. Farmers, struggling with drought and a weak export market, abandoned thirsty crops like citrus and cotton that could no longer be justified, and some shifted to relatively drought-tolerant crops such as sunflowers, or to yet more capital-intensive methods of production such as hothouses.

In 1991, threatened with a 40-percent cut in their allocation of water, farmers did not petition for a restoration of this allocation but rather for monetary compensation for the losses they would suffer as a result. This is an interesting response, and in a way, a suggestive one. What it says is that there are other ways of subsidizing agriculture than providing cheap water and that agriculture can do with less water if it is provided these alternative forms of subsidy.

Nevertheless, the government has applied little energy or creative attention to developing alternatives, just as it has failed to develop a consistent plan to withdraw water from the agricultural sector. The year 1992 brought unprecedented rains, flooding the cities and countryside. While it was acknowledged in all sectors that these floods would not solve the long-term water deficit, they discouraged serious attention to the problem once again.

Criticisms of the way that water was allocated and managed in Israel have been extant in the literature on this subject since at least the mid 1970s, and recommendations for alternative systems that now seem attractive to the national authorities had been urged by various experts for two decades (Shuval 1980; Vardi 1980; Nativ and Issar 1988). Basically, these consist of two different and opposing, though not mutually exclusive, approaches. One is a technical approach, which is costly: produce more water through nonconventional means. The most obvious, treating urban waste water (Selbst 1980), was indeed part of the comprehensive national plan, and has already been put into effect for decades, though not at the rate planned or necessary. Desalination, while also obvious, was and continues to be controversial because of its high cost. Mining of archaic aquifers (Issar and Nativ 1988), on the other hand, has generally been dismissed as an alternative for a variety of economic, technical, and ideological reasons. Even importation of water from other lands such as Turkey or Egypt has been considered, though also dismissed, for similar reasons.

The second, opposing approach is an economic one. This is to decrease the demand for water and increase efficiency of use by pricing mechanisms. Most common is the recommendation that farmers be obliged to pay the true price of the water they receive. This means that water would not be wasted by being used for unprofitable crops or in areas where water is too costly to deliver. Consequently, water would be used much less in agriculture because today, most irrigated crops grown in most areas bring prices too low to warrant the investment. Of course, this would mean a dramatic reduction in the number of small farms and their elimination in a number of regions of the country.

## CONCLUSION

By 1986, agriculture's allocation of water had fallen from 80 percent to 60 percent of the total produced (see Figure 4.3). Not only were farmers using far less water than before, but increasing numbers were dropping out of farming for financial and other reasons. Many of the younger generation did not follow their parents into farming—either it was not economically attractive, or it was not professionally or socially attractive. The older generation was beginning to retire, and because of sharply dropping land values, many found it impossible, given the debt crisis, to sell their holdings. While formally the land distribution system did not change, in fact, there was a shift toward consolidation of farming practices: Fewer farmers were farming more land, and more land-holders were farming less. Many moshavnikim had adopted greenhouse production, especially of flowers, for export. The impracticality of attempting to make a living on a small plot in the modern world was beginning to catch up with reality long before formal decisions were made by government policy makers. As several chapters in this book point out, the central part of the country had become so urbanized that many former agricultural communities were now suburbs with little farming. In other words, the process of adjustment to a variety of economic and demographic processes was already taking place. These adjustments would reduce the impact of a change in the allocation of water to agriculture.

The transformation of farming from a way of life to a type of business has prepared the ground for a substantial policy shift in the treatment of water as a factor of production. It is far easier now than it might have been twenty or thirty years ago to consider charging farmers the true cost of water production, to treat farming as another type of industry. Farming has lost its sanctity, and water, no longer a means of making a nationalist dream come true, is no longer part of a sacred mission but a business input or a domestic utility. There is no sense in asking urban dwellers to sacrifice for the sake of a tiny group of small businessmen. The demise of the moshav, at least as a socialist entity, has made of small farmers, small businessmen.

Had there been no drought in the 1980s, there would still have been a water crisis, for this was predicted by virtually all experts by the 1970s in view of the natural constraints on supply and the failure of the national water management establishment to find ways of reducing levels of consumption. The droughts of the 1980s and the great influx of new immigrants from the Soviet Union merely accelerated awareness of an inevitable process. At the same time, other processes were in place that suggested the direction of response by government authorities and by farmers. These processes pointed in the direction of a reduction of support for farming subsidies, reduction in the allocation of water to farming, changes in the methods of farming and the organization of farming, and dramatic increases in the prices of water in accordance with decreasing availability. Changes in the political and ideological climate of farming have apparently been a necessary precondition for government authorities to play the regulatory role that was called for by critics more than a decade earlier. These changes centrally involve a shift from socialism to capitalism as guiding principles for management of the rural sector.

## REFERENCES

Galnoor, I. 1980. "Water Policymaking in Israel." In *Water Quality Management under Conditions of Scarcity*, H. Shuval, ed., pp. 287–313. New York: Academic Press.

Issar, A., and R. Nativ. 1988. "Water Beneath Deserts: Keys to the Past, a Resource for the Future." *Episodes* 11(4): 256–262.

Kramer, M. 1987. *Three Farms: Making Milk, Meat, and Money from the American Soil.* Cambridge, Mass.: Harvard University Press.

Laster, R. 1980. "Legal Aspects of Water Quality Management in Israeli." In *Water Quality Management under Conditions of Scarcity*, H. Shuval, ed., pp. 263–285. New York: Academic Press.

Nativ, R., and A. Issar. 1988. "Problems of an Over-Developed Water System: The Israeli Case." *Water Quality Bulletin* (13)4: 126–132.

Sadan, E., and S. Pohoryles. 1979. "Patterns of Agricultural Adjustment in an Arid Ecosystem." In *Arid Zone Settlement Planning*, G. Golany, ed., pp. 292–326. New York: Pergamon Press.

Selbst, N. 1980. "Economic, Social, and Administrative Considerations in Reuse of Wastewater." In *Water Quality Management under Conditions of Scarcity*, H. Shuval, ed., pp. 243–262. New York: Academic Press.

Sexton, R. 1990. Perspectives on the Middle East Water Crisis: Analyzing Water Scarcity Problems in Jordan and Israel. ODI/IIMI Irrigation Management Network Paper 90/3f.

Shuval, H. 1980. "Conclusion: The Impending Water Crisis." In *Water Quality Management under Conditions of Scarcity*, H. Shuval, ed., pp. 315–337. New York: Academic Press.

Vardi, Y. 1980. "National Water Resources Planning and Development in Israel—The Endangered Resource." In *Water Quality Management under Conditions of Scarcity*, H. Shuval, ed., pp. 37–49. New York: Academic Press.

# PART II

# COMMUNITIES IN CHANGE

## Chapter 5

# Looking Backward with a Forward View: Stories That Make Sense

*Esther Schely-Newman*

*During the first decade of Israeli statehood, about 300* moshavim *were established, many of them settled by recent immigrants with little or no experience in farming. This chapter examines attitudes of settlers toward the establishment of a new community as those attitudes appear in stories told forty years later. Schely-Newman's purpose is to show how personal narratives explain past events and reaffirm ties between people and place.*

A moshav is an Israeli agricultural community combining cooperative elements with individual family farms. The ideological basis stresses principles of self-work, reciprocal guarantees, and communal help. The farms are not owned by the settlers but are established on state land. Membership in the moshav is decided by the community and acceptance of new settlers is subject to the approval of the group (Weintraub, Lissak, and Azmon 1969). The principles guiding moshav organization require an ideological predisposition and acceptance of communal values. These principles go hand in hand with the values of pioneering Zionism of the Second and Third *Aliyot* which stress manual labor, particularly farming, as an ideological endeavor.[1] Tilling the soil is a means for redeeming the land of Israel and making it Jewish. At the same time, work, particularly farming, redeems the Jewish people from their *galut* (exile) mentality, making them into productive members of society.

The Israeli declaration of independence (May 1948) was followed by mass immigration, doubling the Jewish population within eighteen months. Settling immigrants in moshavim met several needs: (1) food production for the growing population of the country; (2) housing and occupation for those immigrants who settled them; and (3) the moshavim also helped scatter the population and reinforce the borders of the new state. In addition, because of the ideological value of farming, agricultural settlement had a social purpose. There was a conscious attempt to reeducate the newcomers, particularly those immigrants from the Orient who had brought with them ways of life, attitudes, thoughts, and emotional reactions vastly different from our own (Weitz 1971). Farming was the method because "there is no doubt that farming as a way of life is the most suitable framework for education and forging a new character" (Gilad 1972).[2] Settlers in new moshavim had to adjust to a new way of life, to farming, and to cooperation—in addition to adjusting to the new country, language, and culture.

Settling immigrants in moshavim was fraught with anxieties because they were seen as lacking the necessary ideological background. The moshav framework was adjusted to meet the needs of recent immigrants with no ideological preparation. Some of the responsibilities regarding means of production were taken away from the settlers and given to outside agencies, such as counselors or regional organizations (Weintraub, Lissak, and Azmon 1969; Willner 1969). The new type of moshav was now called a *moshav 'olim,* "immigrant moshav," to be distinguished from the "veteran" *moshav 'ovdim.* To assist the immigrants in the moshav, particularly those lacking a "predisposition to change" (Weintraub and Lissak 1964), a *madrikh hevrati* (social counselor) was assigned. The counselor's role was to help the immigrants shed their old values and culture and acquire the desirable, positive, modern, Western–Israeli culture—a process of desocialization and resocialization (Bar Yosef 1968).

This situation created dependence and a feeling of powerlessness on the part of the immigrants who became objects manipulated by the settling authorities. The assumption was that settlers with a lesser disposition for change would have greater difficulties in adjusting to new farming life. New immigrants who were brought unprepared to newly founded moshavim but nevertheless settled deserted areas of Israel were considered at best "reluctant pioneers," a term coined by the anthropologist Alex Weingrod in his study of a moshav 'olim in the Negev (Weingrod 1966).

Tsomet, the moshav discussed here, has a slightly different history.[3] The group of immigrants who settled Tsomet was not selected by the authorities but rather organized independently. A number of young men, members of Zionist youth movements in Tunisia who came to Israel during the War of Independence, recruited other Tunisians to form a group that would settle in a moshav of their own. They spoke some Hebrew, were familiar with farming ideology, and were enthusiastic about the venture of settling in the Negev desert. However, the majority of prospective settlers were not exactly sure about what life in a moshav or in the Negev entailed. Itzhak, the main organizer, admits that people came for economic reasons, not knowing about obligations and bylaws of the moshav. Nusha, an articulate woman whose family was among the first recruits, explains, "We thought moshav is a name. There is Tunis and Gabes and Sfax, and there is also Moshav."

When asked about first impressions of KM 107, later to become moshav Tsomet, people provided similar images: it was *sahra, sahra el khla, la hajra u-la sajra!* (desert, desolate desert, no stone nor tree) as Yvette put it. Ruthie, who was a child then, says that even stones to play with were difficult to find. Nusha recalls her first impressions of the place in early 1950:

> Good God! Only piles of sand to the sky, not even a green branch, nothing, nothing. No bird, no green, no car, no road, no houses, nothing! What is this? Where is it from? What is it? And there were eight tents maybe ten, I don't remember exactly.

Not only women had this image of the place; men who were active in organizing the moshav also described the area in similar terms. Moshe claimed, "When I came here it was wilderness, you didn't see a tree, a bird, or a flower." Itzhak stated, "It was empty, a desert."

This bleak image represents not only the actual view of the place but applies to the early stage of community life when people felt dependent on the authorities who decided for them where the community would be and what work the settlers would do. When asked if he knew what was expected, Itzhak answered,

> To tell you the truth, we knew nothing. It was a challenge for us as youngsters to come and settle in such a desert. . . . I was a bit of an idealist and a dreamer. . . . I acted like the Israelites in the desert, first we obey and then we listen.

At the time, however, in December 1949, the knowledge of Hebrew, the level of education, and the idealism of the group of young men did not pass unnoticed. The potential leadership was recognized by the settling movement. Several individuals who stood out, mainly Itzhak (the main organizer) and Barukh, were recruited and groomed by the authorities as leaders of the new moshavim in the Negev. Itzhak was sent by the Jewish Agency back to Tunisia to recruit immigrants, and Barukh became a political figure in the national arena; other young men were sent as *madrikhim* and professional workers to other moshavim established later in the area. The young Zionist immigrants who spoke Hebrew contributed to the making of Tsomet into a showcase, the "jewel in the crown" of the new moshavim in the Negev; an example of a successful attempt to transform immigrants into productive farmers. Tsomet thus was a place for visits by dignitaries, political rallies, and celebrations honoring leaders such as David Ben-Gurion. This success facilitated the allocation of additional funds for the moshav, expressed in the paving of roads and having telephone lines installed earlier than in neighboring moshavim. The particular situation of Tsomet settlers is expressed in the first report filed by Abraham Elati, the madrikh hevrati who followed the settlers from the immigrant camp. He explains that the people came from

> countries that only in the beginning of the twentieth century began to awaken from the feudal slumber of the retarded Ottoman rule. In their social and cultural ways there is an opposition between the influence of the French language and culture learned in schools and the still existing deep roots of nineteenth century social life. This is particularly so for the older generation, over 35–40.[4]

However, there is hope for the group because

> It must be noted that a lot was done by the army, which brought all members under the yoke of free and democratic life from inner joy and will. The influence of the new style of the labor movement, expressed in free army life, is evident everywhere. . . . One can assume that the negative marks of the past will be blurred quickly enough, forging man's character on the anvil of life in this new land.[5]

Individual success and communal esteem notwithstanding, the community at large had to cope with changes in lifestyle and the daily struggle in an alien environment. Everyone, including the leaders, had to adjust to a new profession, farming, to life in a small peripheral settlement, and to cooperative living. More to our point, the settlers had to learn to live in a desert area and to cope with fears of Arab infiltrators.

The positive disposition for change in itself cannot explain how the community overcame the difficulties of adjustment. The question remains, how were people able to make KM 107, the deserted location to which they were sent, into Tsomet, a community with a distinct history and common purpose? We wish to suggest that the physical acts of planting and building were not sufficient to make a location "home." Rather, the ability to formulate past experiences into stories is the process by which this transformation was made possible. These experiences are reconstructed into stories that reflect the sociocultural traditions and values of the settlers. Through the act of narration, the settlers become full participants in the Zionist ethos, no longer *'olim* (immigrants) but *halutzim* (pioneers). This transformation is evident in narratives of the founding generation of the moshav, who describe their initial encounter with the locale of the community as bleak and full of despair (see earlier discussion in this chapter). At the same time, they express pride over what they feel as their own achievements.

MOSHE: There is a pride in what we built . . . we did, we contributed our share.

HENRY AND MAURICE: We were pioneers, and at the end we built the State of Israel.

One type of story about the first years in the community best demonstrates this transformation. Stories about real sacrifices people made—about lives lost in the process of building the community—create different relationships between people and place. Shooting accidents were not uncommon in the history of the newly established moshavei olim (Shokeid 1971; Willner 1969). Our interest, however, is not in the facts but in the explanations given to them. These stories are an important part of the historical discourse in Tsomet; their analysis explains the process of claiming the right to be called pioneers.

## The Facts

In February 1951, a woman was wounded by a sentry. In August 1951, a sentry was killed by another sentry, and in the winter of 1955, two sentries were shot at by a community member. One was killed and the other seriously wounded. These facts are not debated: The graves of the two men exist, and at the time, brief items appeared in newspapers. Police records of the events are not available because records are destroyed after a certain period of time in cases with no public interest. Lack of public interest, however, does not imply lack of private interest: The families of the victims and the community as a whole continue to tell stories about these events. The case of the woman who was shot in 1951 serves as a model of improper female behavior and gender relations and is the basis of a moral tale.[6] Our concern here is narratives about accidents involving male victims that became stories of revival and self-reliance.

Claude, the first victim, was killed in August 1951. He and his fellow sentry finished their shift guarding the central shed where all farm animals were kept for protection. After the other sentries took over the shift, Claude walked off to urinate. The sentry on duty saw a figure in the dark, called for a password but was not answered. Claude was wearing a white head covering, perhaps a towel, that caused the sentry to believe it was an Arab wearing a *keffiya*.[7] The sentry shot at the suspicious figure and killed his comrade Claude.

According to the newspaper account, the police suspected foul play because of the recent elections (July 1951). Apparently, as some narrators also point out, Claude was a member of the Zionism Klaliyim party and not a member of MAPAI like most people in Tsomet. In the end, the police were convinced that the shooting was accidental and closed the case. Claude was the first death in Tsomet, and thus no previous arrangements were made for a graveyard. It was left to the community and to Itzhak, then the *mazkir* (moshav secretary), to make the decision about where to bury the dead. The place chosen was a small hill adjacent to Tsomet. Itzhak explains his reasons: Claude worked in the vineyard and the hill overlooks his working place, and, in addition, the hill is close to the wadi and is not suitable for cultivation.[8] The communal synagogue that had just been completed was named after Claude.

Nineteen years later, on a Saturday afternoon, the roof of that synagogue collapsed. That same day, Gadi, a young soldier from Tsomet, was killed at the Suez Canal during an artillery barrage from Egyptian troops.[9] The new synagogue built to replace the older one is named after Gadi.

The second case of accidental killing happened in the winter of 1955. Dov was returning home from a family celebration with his wife and child when he noticed two men walking between the houses. It is not clear whether Dov called for the password, but, in any event, the two men remained silent and Dov opened fire, fearing they were Arab infiltrators. The two men were Atiya and

Ralph, the sentries on duty that night, who had stopped for a cup of coffee at a friend's house and were on their way back to their guard posts. Atiya was killed and Ralph seriously wounded in the accident.

There were other incidents of shootings, though without personal injuries. Reuven tells about his fifteen-year-old daughter, Gizelle, who shot at stockings hanging on a laundry line, thinking they were a person. Roger tells of an incident of a stolen cow, returning to Tsomet on its own, shot by a sentry who thought it was an Arab. In general, Tura remembered, "Shots were heard every night."

It is interesting to note that people insist they never feared Arabs in Israel. The inconsistency between explicit utterances and actual behavior exists in many narratives relating to the anxiety over security in the early years. Some people blame the shooting on lack of training in use of firearms. Moshe explains, "The problem was that people were under pressure and were not trained as they are today, or else they would not shoot before hearing an answer."

RIVKA: It is because of the pressure and fear. So people pull the trigger too quickly and that's what happened.

ITZHAK: There was a lot of tension and shooting that resulted from nervousness.

JULA: They gave us guns and there were people who . . . they did not know how and they were afraid. With fear they would hold the Sten [gun] and shoot wherever it will hit, wherever.

For the people of Tsomet, the explanation that some individuals were not careful in using guns and were afraid of Arabs is not sufficient to explain the tragedies. Accidents are difficult to accept because they are arbitrary. Traditional societies tend to search for causality, relationships between occurrences or explanations that reassure people life is rule governed and events fit a larger scheme of things. The function of the stories about the killings of Claude and Atiya is to explain them according to traditional values and in terms of national revival.

### The Narratives: Claude's Death

In telling about Claude's death, people never fail to mention that it happened on the night of *Tisha b'Av* (the ninth day of the Hebrew month Av), the date on which the Temple in Jerusalem was twice destroyed. This day culminates three weeks of mourning commemorating the end of the independent kingdom of Israel in 70 C.E. Tunisian Jews consider this three-week period as dangerous because individuals are susceptible to tragedy, particularly on the night of Tisha b'Av. The violent death of a member on that night could be interpreted as a bad omen for the future, for the attempt to revive the deserted land. In the narratives, the death of Claude is perceived as a sign of destruction and of self-redemption. That is, he died on the night that symbolizes destruction, but he died protecting his homeland. "Our first friend Claude *nafal korban bashmira* [fell victim while guarding]," using a term reserved usually to describe soldiers who fell in the line of duty. The interpretation of the accident as death at the service of the community is reinforced by Itzhak's ruminations about the first grave. Itzhak said that despite the sadness he felt, nevertheless, that death is a step in the process of nation-building. He quotes a line from a poem of Alphonse Lamartin: *"Ce sont les cendres des morts qui créa la patrie"* (The ashes of the dead create the fatherland). His view suggests that there is no growth without victims and that death is needed for the (re)birth of the place. Thus, the first death in Tsomet, not natural but a violent one, may be seen as a covenant with the land that requires its share of blood in order to let the newcomers plant roots in it.[10] Tunisian tradition gives honor to scholars and rabbis in religious ceremonies by naming synagogues after them. In naming Tsomet synagogue after Claude, the community has extended the traditional honor given to scholars to acknowledge his sacrifice for the communal good.[11]

The death of nineteen-year-old Gadi at the Suez Canal in 1969 also serves as a point of transition from old to new, from protecting the homestead to becoming fully incorporated in the State of Israel by shedding blood for national security. The death of the young Sabra is remembered and memorialized in the traditional way by naming the new synagogue after him. Claude belongs to the old guard. The new generation and new synagogue belong to the Sabras, those born in Israel and dying for it. In both cases, the death and the memorial synagogues reaffirm the connection between destruction and revival, between death and building a new land, and between the individual community and the greater whole, the nation-state. A local version of the national Israeli theme of sacrifice and revival is repeated every year in the commemorative ceremonies of Memorial Day and Independence Day (Handelman 1990).

### The Narratives: Atiya's Death

Atiya's accidental killing is attributed to the mental instability of Dov, who had been institutionalized for many years. Nevertheless, this death also is explained in ways that stress basic communal values. There is a family relation between the two deaths: Claude, who was killed in 1951, had a sister who was married to Dov, the man who killed Atiya in 1955. Such relationships are not rare in a community of immigrants from the same country.[12] The relationship between victim and perpetrator did not pass unnoticed, and Yvette, who is known for her exceptional memory and gift for details, adds an element of revenge to connect the two incidents: "Dov's wife told her husband, she told him if you do not avenge me and kill somebody from Tsomet for me, you will not be my husband and I will not be your wife." And indeed, Dov did kill someone from Tsomet. This story follows the traditional maxim of "everything comes from the wife," a proverb frequently used to explain men's behavior both in positive and in negative instances.

Another story narrated by Yvette offers a quasireligious explanation of the 1955 tragedy. A couple in Tsomet was engaged to be married, but the bridegroom disappeared before the ceremony. The rabbi predicted that a tragedy would occur because the bride had already gone to the *mikveh* (ritual bath) and, therefore, her condition was now similar to a desecrated Torah scroll. According to Yvette, that same night (or week) the shooting occurred, leaving one man dead and one seriously injured. Women who heard Yvette tell this story agreed that a bride deserted before consummating the marriage but after visiting the ritual bath poses a danger to the community. Apparently, the couple did eventually get married but had an unhappy life, a story known to other people in Tsomet.

These two stories narrated by Yvette reiterate the importance of women to the prosperity of the community. Deserting a bride has consequences beyond personal feelings because without marriage there are no children and, consequently, no growth in the community. Atiya's death thus can be seen as an atonement for the sin of another man (the bridegroom) against the community.

## DISCUSSION

For the people of Tsomet, Claude, Atiya, and Gadi are heroes, victims of a wartime situation and not of human error. They are symbols of the community in struggle; the lost lives are the price to be paid in the process of building a community and by extension, building the nation. The stories, therefore, serve the community in achieving self-esteem and creating an esprit de corps. These death narratives reaffirm the settlers' attachment to the land, the roots planted in blood, combining traditional beliefs and national myths. The narratives stress the independence of the community: The police are not involved and their assumptions of political motives for murder are irrelevant to the settlers. What the settlers seek and find in the narratives are internal explanations that make sense according to their own values. Even if the stories do not exactly mirror the events of the past, they nevertheless reveal the truth of experience. By rearranging historical events into meaningful stories with cause-and-effect relations, people are able to explain the events to themselves (White 1981). Such explanations give the settlers the right to claim the place as their own. The victims are local members who died for a communal cause. Protection of and dying for the homestead are not acts of the authorities, the Jewish Agency, or the madrikhim. Even if they died as a result of errors committed by fellow settlers, the victims are regarded as heroes sacrificed for the common good.

Religious explanations are still evident today. In 1988, a young man was killed in a road accident. During a conversation with Jula in October 1991 about the lack of morals in modern times, she complained that the local rabbi is not doing enough for the community. She cited a case where she felt that the rabbi should have forced a married woman, who was having an affair with another man and whose behavior was known to others, to be divorced, thus removing a curse from the community. This curse, Jula explained, already cost the life of the young man killed in the accident.

The involvement of women in the stories also marks the transition from a frontier post to a community. Claude died when he was patrolling the area around the shed where livestock were kept; he was killed by another man performing the same duty. Atiya's death four years later occurred in a different situation. Atiya was guarding the houses where women and children sleep; he was killed by a man walking with a wife and child. Frontier life was over by 1955. Women and children now take precedence over livestock. Women make the frontier post into a community: They are bearers of the following generations. Women have the power to build (bear children) but also to destroy, as they may influence men to kill or cause misfortune by their behavior. This may be the epitome of *'arevut hadadit* (reciprocal guarantee): Moshav members are not only responsible for each other's fiscal situation but also for their morals. Therefore, sinners must be chastised. The behavior of individuals affects the others, and only proper behavior of all can ensure the well-being and security of the group.

## CONCLUSION

An employee of the Jewish Agency told me that at the time, "reality dictated doing things from above," meaning that there was no time to ask the new immigrants what was the best way to go about introducing change.[13] This widely held opinion explains the treatment of new immigrants, particularly those from non-European countries, during the first years of Israeli independence. The immigrants were seen as objects to be manipulated by the authorities—the settling movements, the madrikhim, and the political parties. Immigrants, it was believed, were not able to make choices, and responsibility was not given to them. Even in a community such as Tsomet where people were given relatively greater choice, important decisions were made by outsiders—where the community would be built, how to name it, what crops to grow, and so on. However, when we listen to the immigrants' views about their experiences, a different picture is presented. In stories narrated by Tsomet settlers, an alternative truth emerges, another set of relationships between the authorities and the newcomers is posited. In their version of the truth, the settlers made their own decisions. It is because of their actions that the community is prosperous. It is not surprising, therefore, that the authorities hardly appear in the oral history shared by Tsomet members. This is clear in the narratives discussed here. There is no mention of officials or police in the case of Atiya's killing, and the suspicion of foul play in Claude's death is quickly dismissed. The decisions about the location of the graveyard and naming the synagogues are made by the settlers themselves. Obviously, Tsomet people are aware that the Jewish Agency was involved and that madrikhim were helping them—we do hear about them on occasion—but their role in narratives is negligible. In their oral history, Tsomet people do not see themselves as objects of manipulation but as pioneers proud of their efforts. As masters of their stories, they become masters of their history and fate.

The leaders of the group say today that it was difficult, but those who wanted to accept the difficulties were able to overcome them and to adjust; others gave up and left. Difficulties of adjustment were part of the lives of people who came ideologically prepared for the new way of life as much as they were part of life of nonprepared settlers. Reluctant or ready for the pioneering endeavor, all immigrants were subject to a difficult process of adjustment. From the stories of the settlers in moshav Tsomet, we may conclude that the difficulties are better understood and alleviated by the act of narration. By looking backward, in telling stories about their communal past, the older generation of Tsomet is able to leave a legacy to the next generation. Narration is a "definitional act," it is a way of portraying the self and the community and of reinforcing communal values (Myerhoff 1986). These stories explain past events in a manner that makes sense for the second and third generations of Tunisian-Israelis who by hearing the stories vicariously participate in building the community and thus feel part of the Israeli ethos. Even when they tell stories that may present them as fearful, the older generation still transmits communal values and a moral *'arevut hadadit* on which the community stands. The "negative marks of the past," for Tsomet's settlers, has not been "blurred away" as predicted in Elati's report; instead, it has been reformulated and merged into the communal history, and thereby making it understandable and acceptable to the founders and a source of pride and belonging to the younger generation.

## NOTES

1. *Aliyot,* waves of Jewish Zionist immigration to Eretz Israel from 1881 to 1948, are numbered; the dates of each are based on political events and changes in the condition of immigration to Eretz Israel. The Second (1904–1914) and Third (1919–1923) Aliyot were vital in shaping pioneering Zionism. Many immigrants were members of the Jewish Socialist party, Poalei Zion (established 1903 in Russia) which combined socialism with national Jewish revival. Poalei Zion and Hapoel Hatzair (established 1906 in Israel) merged and split several times to create MAPAI, now the Labor Movement. For a discussion on the hegemony of pioneering Zionist ideology, see Vital (1982, pp. 390–411).

2. Rubin's (1959) anthology commemorating the first decade of moshavei 'olim includes numerous references to the task of the settlement authorities in reshaping the personality of the immigrants.

3. Tsomet was settled by Tnu'at Hamoshavim (settlement movement of MAPAI) in December 1949 by new immigrants from Tunisia, many of whom still live there. It was one of the earliest moshavei 'olim in the western Negev. The data for this chapter were collected between 1987 and 1990 for a doctoral thesis treating oral narratives about Tsomet in the formative years, 1949 to 1956. Past and present residents of Tsomet and officials involved in establishing the community were interviewed. Names have been changed to protect people's privacy. I have translated all quotations from Arabic and Hebrew (Schely-Newman 1991).

4. Letter to Tnu'at Hamoshavim, dated February 9, 1950, two months after the first settlers arrived (the movement's archives).

5. Ibid.

6. This event and its narratives were treated elsewhere; see Schely-Newman (1991, 1993).

7. One narrator, Moshe, said that Claude was wearing a red *keffiya.* The color of the *keffiya* seems to justify the shooting because the checkered red and white head cover became the symbol of the PLO and its supporters. The question of plausibility—was it possible to detect the color at night and not recognize the person?—is not an issue for the narrator.

8. For Tsomet people, the wadi represents the remaining unconquered wilderness: the source of evil, where Arabs and jackals still roam. The choice of a place adjacent to the wadi for a graveyard fits this perception (Schely-Newman 1991).

9. The ongoing artillery exchange between Israeli and Egyptian soldiers stationed on the opposite sides of the Suez Canal from 1968 to 1969 is known as The War of Attrition.

10. The interpretation of death as blood covenant between new immigrants and the land was raised in February 1992 after the killing of two Russian immigrant soldiers and the knifing of two women in the town of Kfar Saba. In May 1992, after a murder in Kfar Darom, a fellow settler was quoted saying, "There is nothing that ties you to a place more than a grave" (*Hadashot* daily paper, May 28, 1992).

11. In his discussion on adjusting religious traditions of south Tunisians in Israel, Shlomo Deshen describes a case of commemoration services to two men who died trying to save a young girl from drowning in a flash flood. Though these men were not part of the community and were not religious scholars, their self-sacrifice was acknowledged and they were given honors usually reserved for scholars and religious leaders (Shokeid and Deshen 1977).

12. In fact, marriage within the extended family or at least between people from the same hometown was seen as desirable among the immigrants. Several groups of related individuals reside in Tsomet.

13. Interview with Micha Talmon, former director of the Settlement Department of the Jewish Agency in the Negev, March 1990.

## REFERENCES

Bar Yosef (Weiss), R. 1968. "Desocialization and Resocialization: The Adjustment Process of Immigrants." *International Migration Review* 11(6): 27–45.

Gilad, G. 1972. *Milestones in Absorption of Immigrants in Agriculture*. Tel Aviv: Center for Education and Culture (Hebrew).

Handelman, D. 1990. *Models and Mirrors*. Cambridge, Mass.: Cambridge University Press.

Myerhoff, B. 1986. "Life, Not Death in Venice." In *The Anthropology of Experience*, V. Turner and E. Bruner, eds., pp. 261–286. Chicago: University of Illinois Press.

Rubin, Y., ed. 1959. *A Decade of Moshavei Olim: An Anthology*. Tel Aviv: Tenuat Hamoshavim.

Schely-Newman, E. 1991. "Self and Community in Historical Narratives: Tunisian Immigrants in an Israeli Moshav." Ph.D. diss. Chicago: University of Chicago.

———. 1993. "The Woman Who Was Shot: A Communal Tale." *Journal of American Folklore* 106: 285–304.

Shokeid, M. 1971. *The Dual Heritage: Immigrants from the Atlas Mountains in an Israeli Village*. Manchester: Manchester University Press.

Shokeid, M., and S. Deshen. 1977. *The Generation of Transition*. Jerusalem: Yad Ben Zvi (Hebrew).

Vital, D. 1982. *Zionism, the Formative Years*. Oxford: Oxford University Press.

Weingrod, A. 1966. *Reluctant Pioneers*. Ithaca: Cornell University Press.

Weintraub, D., and M. Lissak. 1964. "The Moshav and the Absorption of Immigration." In *Agricultural Planning and Village Community in Israel,* J. J. Ben-David, ed., pp. 96–101. UNESCO.

Weintraub, D., M. Lissak, and Y. Azmon. 1969. *Moshava, Kibbutz, Moshav.* Ithaca: Cornell University Press.

Weitz, R. 1971. *From Peasant to Farmer: A Revolutionary Strategy for Development.* New York: Columbia University Press.

White, H. 1981. "The Value of Narrativity." In *On Narrative,* W. J. T. Mitchell, ed., pp. 1–23. Chicago: University of Chicago Press.

Willner, D. 1969. *Nation-Building and Community in Israel.* Princeton: Princeton University Press.

Chapter 6

# Change or Renewal? Israel's First Moshav Ovdim in a New Political and Economic Order

*Naomi Nevo*

*Because Nahalal was one of the original successful moshavim and for so long played a leadership role in the moshav movement, what happens here is of special interest to everyone connected with the moshav. Naomi Nevo, who wrote a monograph on the first moshav ovdim based on her study of the community in Nahalal, returned to look at how the members adjusted to recent changes in the economic climate. She found that while there were some changes, the pattern of adjustment followed a template designed in Nahalal's past, with the same differences of approach and opinion pursued even by the same member families.*

The establishment in 1921 of Nahalal, the first moshav ovdim, was the culmination of many years of discussion and polemics among the Jews who had immigrated to Palestine with the Second Aliyah[1] looking for an alternative to the kvutza[2] way of life that would fulfil their economic and social expectations. The itinerant agricultural workers together with some of the pioneers who had helped establish Degania as the first kvutza set up an organization to implement the ideas of Eliezer Joffe[3] as formulated in his blueprint for a cooperative small holders' village.

Unlike the kvutza, the moshav came into being with a philosophy and ideology already formally and explicitly spelled out in response to the then prevailing conditions of dire poverty, hardship, and danger. The raison d'être of the moshav was to ensure the supremacy of the nuclear family both in the domestic and in the productive sphere. The family was to be contained within a community framework of cooperative and exclusive economic and social institutions. This meant that every member of the moshav had to be a permanent resident in the village, fully participating in its social and economic life.

Founded on nationally owned land, Nahalal's farms began with plots that were equal in size and quality. Their continuing development, however, depended on the talents, entrepreneurship, and work (hired labor was prohibited) of the farming families themselves. Community guarantee of loans and a system of mutual help were instituted to bolster the individual farmer in need. Global responsibility of the community for health and education extended that principle, thereby ensuring the true egalitarianism of a righteous society. Marketing—purchasing the means of production as well as selling the product—was carried out by Nahalal as a consumers' and producers' cooperative, thus freeing the individual farmer from the stigma and unpleasantness of trade and bargaining, as well as strengthening his position vis-à-vis external economic institutions. In short, the moshav was to act as a corporate entity (known as the aguda) so that it could provide a supporting framework for the individual farming families. The ideology of the moshav, then, was formulated to meet the needs, as well as the social aspirations, of its founders in that particular period.

There were, however, two philosophical sources from which the founders drew their inspiration—socialism and individualism. The dissonance inherent in this dichotomy has plagued Nahalal throughout its history. In any case, differentiation was incipient in both the economic and occupational spheres of the new society. An extreme example was the differentiation between farming and service members—the former earning their living from their farmsteads, while the latter received wages from the aguda for various technical and other services. There were even some independent craftsmen, but these eventually became employees. Although they were members of the moshav with voting rights, the service workers were not accorded rights in property. They were permitted to live in the village only as long as they worked there. As for the farmers themselves, despite the arrangement that farm plots were qualitatively and quantitatively equal (even equidistant from the village center, which dictated the circular plan of Nahalal)[4] right from the beginning, some farmers came with capital in the form of money or livestock, while others came with nothing. This clashed with one of the main planks in the moshav platform—initial equality of basic resources, land, water, and a modicum of capital. Already present in the early years of Nahalal's subsistence economy, this inequality became more persistent in the later years of a developing market economy and modern technology. Parallel to this process, possibly even because of it, the role of the moshav in its corporate entity of the aguda crystallized in the productive sphere, resulting in the multiplication of constraints on individual family enterprises.

Risking oversimplification, I nevertheless make the point that the members who advocated more corporate control over the individual farmer's economic and social behavior—subscribing to the concept of the Collective Will—eventually demanded legal enactment of this principle. The proposed law would have empowered the moshav management committees to enforce compulsory regulations of membership as defined by the Moshav Movement (see Chapter 1). Any individuals and their families who were judged by a special court to be nonconformists could then have been legally expelled from the village. The moshav law was never passed in the Knesset, but the proposal split the Nahalal community into pro-law and rebel factions. The former were identified with the establishment both of Nahalal and of the Moshav Movement, presenting themselves as the orthodox moshav ideologists, while the latter were proud to call themselves nonconformists but purists as well, who claimed to represent the true, unadulterated moshav philosophy of Eliezer Joffe who had abnegated the supremacy of the Collective Will and fought against institutional coercion.

In fact, many (but by no means all) of the rebel faction refused to accept one of the original imperatives of moshav ideology—that the exclusive occupation of its members be farming—and also impinged upon one of its basic norms, that of self-labor. Because they were not full-time farmers, it was more convenient for them to plant orchards rather than to tend the livestock that was considered the most vital component in the mixed farming economy stressed in the moshav value system. Some of this group habitually sold their produce privately, that is, outside of the organized cooperative channels, thereby making it difficult for the aguda to debit them for various payments, including municipal taxes. It is therefore relevant here to emphasize another bone of contention between the two factions, namely, the principle of the moshav entity as a combination of the economic, social, and municipal aspects of the community. Up to 1959, a committee for municipal and social affairs functioned in Nahalal side by side with the committee that managed the farming, marketing, and other economic functions of the cooperative village. The moshav, that is, the aguda, as one indivisible social and economic institution (acting as a credit association for its members) had long since become an accepted principle for all the other cooperative small holder villages in the country. It is significant that their prototype, Nahalal, hesitated for so many years before deciding to implement what has since become recognized as an essential characteristic of the moshav ovdim. But the decision was hardly by consensus even then.

When I first arrived in Nahalal at the end of 1971 to start my field work (which was to eventually culminate in my doctoral thesis "Social Change in a Veteran *Moshav Ovdim*")[5], the situation between the two factions was one of detente—carefully nurtured by both sides—not least by the pro-law people. After seventeen months of participant observation, I left the village, wrote up my material, and paid only desultory visits to the village over the intervening years.

About eighteen months ago, I returned to Nahalal in an attempt to discover how the first moshav had fared in the aftermath of the economic and social crisis that had so recently hit the agricultural system of this country (see Chapter 13). To tell the truth, I found it difficult to classify the changes that occurred since I had completed my intensive field work some seventeen years before. Which of them had resulted directly from the crisis? Perhaps the crisis had merely sharpened that historic conflict that I have already described as having characterized Nahalal for so many years; that ever present chasm between the ideologists and the nonconformists.

True, the tacit agreement (so well known to anthropologists) to avoid official recognition of ideological sin and so obviate the need for action had—with only a few dramatic exceptions—been working well for over twenty-five years. Now there was an important difference. Nonconformism, present from the dawn of Nahalal's existence in 1921, was being legitimized. From the very beginning, the *Nahalalim*[6] fought among themselves over all their ideological concerns of yesteryears—the Collective Will, the practice of institutional control, and, particularly, the principle of global guarantees that mean every member is responsible for every other member, as well as for the economic institutions of the community. All these principles were to become highly relevant to the changes of later years.

However, during that visit a year and a half ago, as well as on my more recent visits, I walked around the famous circle, savoring the familiar rather than puzzling over contexts of change. I felt that I was walking on firm ground as I discovered that on the great majority of farms the same families were occupying the same farmsteads, dramatic evidence of the continuing line of descendants. While the Nahalal families are still accorded prestige and honor (even in the face of the 1977 right-wing Likud victory that occurred for the first time in Israel's political history), they are, nevertheless, mere mortals. This fact enables the funerals to continue as the most significant ritual renewing the solidarity of Nahalal as a community.

"We Nahalalim do not divorce; we commit suicide," one famous farmer remarked to me. During the last eighteen years, only one couple has divorced, but they still occupy the same farm together (there is no legal way of dividing a farm in Israel). Another four Nahalalim have committed suicide, thus joining those who in previous years chose this particular way to leave Nahalal. I do not intend this remark as dark humor. It comes from a particular context. Many Nahalalim have told me over the years how difficult it is to leave the village, and they include some who managed their exit and even their entrance to another moshav successfully. Their mixed feelings were expressed by one in the cliché of a "love–hate relationship." Not one of the suicides of recent years had any connection to personal economic tragedies as in some other moshavim.[7]

So round and round the famous circle I went, and I began to feel dizzy with so much more of the same. I didn't know about Nahalal, but *I* certainly needed a change and I began to look for it. I recalled that a farmer's wife had once opened a boutique in her farmyard. It hadn't lasted long in the then prevailing ideological climate, but now I suddenly noticed a studio for ballet dancing (housed in an unused hatchery), a furniture shop, and a couple of artisan workshops. All these, unlike the previous, almost clandestine, nonagricultural enterprises, the owners of which were subjected to social ostracism, are situated in the farmyards and accepted and patronized by the Nahalalim, as well as by clients from the surrounding area. Some agricultural activity was also being advertised to attract the private consumer, and I read the signs at the entrance to the farms that said "Fanya's Nursery" or "Buy your cuttings from Chava."

I walked on, encircling the village and so returning to my starting point—the center, or the "Acropolis" as the founding fathers liked to call it. Here, the economic and social institutions had stood side by side, their close proximity fulfilling the visions of the moshav's planners. The cooperative grocery, the dairy, the hatchery, the storeroom, and cold storage plant did not disturb the school, the *Beit Ha'am* (assembly hall and community center), the secretariat, the clinic, or the synagogue. But during my sojourn in the village twenty years ago, the economic institutions had already dwarfed in every way all the other functions of the Acropolis. The old-timers complained bitterly that the Acropolis had become a marketplace. And indeed, the turkey slaughterhouse exhaled its stench over the community center garden, while the flying feathers and even discarded parts of the birds were deposited for garbage disposal in the amphitheater. And over all, the monster silo sprayed its dust as it prepared feed for the livestock.

I saw that the old vine-covered schoolhouse had been replaced by a large utilitarian building, which is now the primary school for the whole district. Nahalal's children, I thought to myself, can no longer needle Kfar Baruch's children, who were once the only other pupils there. The pupils now come from many other settlements and do not necessarily appreciate Nahalal's special historic position. Nahalal is no longer the seat of the regional council although the wooden huts remain to serve as modest offices for equally modest agro-technical local functions managed—until his recent untimely death—by a second generation ideologically orthodox Nahalal farmer.

As I stood there, the tranquility began to disturb me. Only the cooperative grocery and the secretariat showed signs of life, and these were somewhat subdued. True, the school had closed for the day, but I was used to seeing a continuous stream of farmers visiting the secretariat at that time of day. There was obviously no longer quite the same kind of relationship between the farmers and the "governors." The amphitheater was as clean as a whistle, there was no sign of a turkey—dead or alive—and the silo stood like a stark, silent monument reaching upwards through the trees. But the really dramatic change I sensed was when I saw the strangers: Arabs, who had rarely been seen in Nahalal in the past, were now going about their obviously legitimate business. Nahalal had always observed the Zionist socialist prohibition against employing Arabs (indeed any non-Jewish labor), even after the Six Day War when it became a norm throughout Israel. With the exception of construction workers (overlooked because they were actually employed by Jewish building contractors and it was only with the latter that Nahalal had direct dealings), Arabs had never been employed on the farms of Nahalal, not even by the rebels. Now I was told, and certainly not in a whisper, Arabs were working on the farms, and Arab merchants were solving the new problem of Nahalal's surplus milk production by purchasing it directly from each remaining dairy farmer. Although still declining in number, the dairy farmers consistently enlarged their milch herds just as they had twenty years ago. Long before and after my own experience of Nahalal, milk production was then considered the farmer's bank because in spite of quotas,[8] no milk remained unsold and rarely were the producers seriously penalized. In those days, prices were controlled, guaranteed by the government, and did not fluctuate. The ideologist pro-Law faction informant who assured me that milk could never be sold outside Nahalal's cooperative marketing system and would therefore be the one certain source for the return of money to the aguda must surely be turning in his grave. One of the results of Arab agricultural labor has been the reemergence of vegetable growing. The branch vanished completely from the village twenty years ago when family labor power was replaced by capital. Now that capital is scarce and Arab labor is available, the vegetables have returned.

What was happening in the service workers' circle? Perhaps only a Nahalali (although possibly also an anthropologist) can appreciate the awe-inspiring event of a service worker's son acquiring a farm through purchase rather than by marrying the farmer's daughter. Perhaps even more significantly, five houses in the center are now up for sale—to be offered on the open market with no precondition that the purchaser be a service worker in Nahalal. Farmers, however, cannot sell their property on the open market. Any potential purchaser has to be accepted for membership by the aguda.

For some years, service workers have been able to register as property owners and so bequeath their homes to their children even though they may be working outside the village. The tied cottage, so infamous in Britain's rural history, has finally disappeared even from Nahalal, much to the relief of some of its farming families who expressed in conversation with me their resolution to correct what they saw as flagrant discrimination.

The generations move on and, of course, out. With all due respect to our somewhat obsessional preoccupation with the social and economic changes incurred through the collapse of the purchasing organizations, (see Chapter 13), the changing of the guard in Nahalal which has taken place is largely due to natural causes. The third generation now governs from the Acropolis. The grandsons of the founders (no granddaughters as yet) make up the establishment and its opposition nowadays. The field work of the last few months has revealed a most fascinating finding, namely, that the differences of opinion among the new leadership follow the pattern of their fathers' factions if not their grandfathers'.

The rebels of the 1960s who planted deciduous fruit trees, who worked outside the moshav, and who marketed privately have sons who fervently believe in free and private enterprise. Sons and fathers of this faction are opposing sons and fathers of the historical pro-law ideologists' faction. But we must remind ourselves that the faction who were the rebels of yesterday have the blessing of the national political establishment of today. Their legitimacy is now unquestioned. Private marketing in Nahalal, while still inconvenient for the management committee, is now permissible. If we look at it as a facility for preventing at least immediate poverty (and Nahalal was always concerned with camouflaging poverty) in a period when it is no longer acceptable to allow members to draw a minimum monetary allowance debited to a farm already debt-ridden, then we must say that private marketing has been granted ideological dispensation. Today, people may jingle coins in their pockets without fear of ambush or the uprooting of their orchards.

It is important at this point to describe the dynamics of Nahalal society after the crash of the purchasing organizations. Up to 1985, Nahalal's financial situation had been excellent. Even during the period of inflation, Nahalal prospered, mainly due to the government subsidies then still given to the dairy and meat enterprises. When the Purchasing Organization of the Northern Region (to which Nahalal belonged) collapsed, Nahalal was caught with a debt of four to five million shekels. An alternative version, as recounted by officials who were not Nahalalim, is that Nahalal's members were enjoying high interest from ten million shekels they had lent to the Purchasing Organization—until the crash. Her reaction was "to fill the farmyard" (i.e., to buy as much stock as she could on credit). The aguda even then got credit because Nahalal's name was good. Always, up to this day, Nahalal has managed to pay its suppliers, but even apart from this obviously very important fact, the name "Nahalal" was in itself still a credit card. Throughout the years of crisis, Nahalal paid its external debts. But internally the situation was going from bad to worse. The members' debts to the aguda grew, private marketing increased considerably, and the situation got out of hand.

Let me dramatize the dry, complicated figures by a narrative. The time is 1987/88, the place Nahalal. During the two months of January and March 1987, thirty-five general meetings were held in Nahalal that about sixty members attended. This in itself was a most unusual phenomenon. After marathon discussions, one of the anti-aguda faction (the politically right-wing) Nahalalim persuaded the assembly to demand internal arrangements as a condition for external arrangements with the banks. Old debts would be canceled, and working capital would be distributed as loans, although only to those whose farms were a good risk and who would return the money through the classic moshav bookkeeping system that depended on organized marketing. Obviously, only those who agreed to market through the aguda would be eligible. Forty-nine members signed the agreement. This meant that twenty-six refused. Eventually, ten of the nonsigners offered to return the loan in cash and their offer was accepted. However, it soon transpired that about twenty-five of the original forty-nine were also marketing outside the system, which, of course, meant that the aguda had no certainty of recovering the loans given in order to provide the farmers with working capital. Indeed, for all practical intents and purposes, the aguda comprised only thirty-four members, a far cry from the Nahalal entity of seventy-five farming families.

What, then, is the status and what are the functions of the aguda today (1991)? First, as all the Nahalalim I speak to remind me, the four to five million shekels debt of Nahalal are the members' shekels. In addition, it transpired that there was an outstanding debt of three million shekels for income tax—the debt of Nahalal's members, of course. It was clear that new arrangements were essential. For example, a silo was built with the money of those farmers who needed fodder. These farmers sold fodder to all from the surrounding area who wished to buy it. Initially, the profits were distributed only to those farmers who had financed the building of the silo for their own use.

However, there was soon no alternative but to distribute the profits, although only from external sales, to all the Nahalalim; "Even the service workers received $500," said my informant resentfully. She added an important detail: The vote in the General Assembly was by secret ballot, which alone constituted a social revolution in Nahalal. The dynamics raced ahead, and today, the silo is managed privately by a man from outside Nahalal who sells fodder to the livestock farmers in the village on a private, individual basis. The aguda, therefore, still formally a credit association for its members, in fact has neither power nor money. For the last eighteen months, there has been no organized marketing through the aguda and, therefore, no credit has been allowed to any farmer.

Since criteria were laid down a few years ago defining the circumstances and conditions under which a member could get loans for purchases of capital equipment (including household furniture and appliances), for farm development and, indeed, all the other needs of a Nahalal farmer, the aguda's management body has no need to request each member to appear personally before it to state his or her case. In fact, the aguda now deals only with the services and institutions belonging to it. But even here its activities have been curtailed. For instance, the garage and its services have been leased to an individual member. It remains with the clubhouses, the grocery, the water system, green fodder, and two or three other enterprises. In most instances, profits are distributed to all the members, including service workers. Today, the aguda no longer controls the land in Nahalal except the roads, a grove, and probably the swimming pool.

At this point in time, we can conclude that Nahalal has suffered severe economic discomfort but no irreparable harm. Debts are indeed enormous, but the farmers are managing to continue without paying them. More economic damage has been caused by the Intifada (the Palestinian uprising in Gaza and the West Bank). The uprising contracted the market for calves since the Arabs in the occupied territories had been their best market. Some of the weak farmers are liquidating the branch, but the strong can manage to keep going. Indeed, this might be a valid comment that is generally pertinent to Nahalal's farmers throughout the moshav's existence. Pragmatic arrangements have always been made to cope with new circumstances, including microcooperative arrangements based on specific farm enterprises in which the strong farmers are engaged. Three turkey breeding cooperatives (one of six families and the others two of three and four families, respectively) were established twelve years ago. There is no doubt (according to the farmers' own remarks) that they were able to "get off the ground," despite the obvious contradiction to official moshav ideology, because of the national political and social atmosphere favoring extrainstitutional initiative. Now, more micro frameworks are appearing, the latest one being a pool of avocado growers who organize irrigation and spraying arrangements together.

Regarding noneconomic adaptations to new circumstances, I have already referred to the fact that Nahalal no longer gives allowances to needy members. Nevertheless, these people are helped considerably simply by the management committee's unspoken agreement not to collect municipal and farming taxes from them: "They own their farms and they can sell them. But in any case we won't let them go under." Listening to the very few Nahalalim who fall into this category, it was absolutely clear that they had no intention of selling their farms and leaving Nahalal.

In conversations with some of Nahalal's third generation during this latest field work, I cannot doubt that they have traveled a long way from their parents. But in so doing, they nevertheless seem to have walked around the circle, arriving eventually at the original starting (or finishing?) point of their grandparents, the founders. The grandchildren's plans for the future of Nahalal were generally in the image of a socially open village community with two separate committees, one for the farmers and one for the urban residents, who will, nevertheless, live side-by-side. The urban residents, however, will be subject to the farmers' rules and regulations. For example, roads and other public property will be owned and controlled by the farmers; restrictions will operate on artisan and other enterprises, as well as on private dwellings, according to their effect on the farmers' comfort and interests. I quote from the remarks of the grandson of an exceptionally prominent founder: "Ordinary citizens will have to realize that they are entering a farming community and will have to live accordingly, even though they will almost certainly be the numerical majority. The people of Israel gave us the land for agriculture."

It made me wonder what had actually changed.

Three months after this paper was presented at the Moshav Seminar held in Sde Boqer, the country was electrified by the news that Nahalal had formally separated its farming interests from the municipal and social sphere of interest. It was seen as the end of Nahalal as a moshav. Impressed by the drama of the media, I rushed to the village, anxious to be with my friends, with those among them who might be sad as with those among them who might be happy. I found neither.

The separation had first been debated at a general meeting when it was proposed by a member who had long ceased his unsuccessful attempts to be a farmer. He has always been considered as rather eccentric and is actually one of the marginal people in the moshav. He had been sure that such a separation would lead to a much more dedicated farm committee and mutual help would once again be implemented. But there were only seven participants in that general meeting (a not uncommon phenomenon), and it was decided to convene another general meeting. That meeting also had to be reconvened since only four participants turned up. The third attempt was rather more successful because eight members were already present at the appointed hour and a few more came later. It was proposed that a referendum be conducted on the question of separating the bodies. Seven voted in favor and three against.

The referendum was conducted very professionally, with a very high turnout but definitely not a total one. Eighty-one voted for separation, while only six voted against. The question remains of why approximately only one-third of the potential voters bothered to vote.

The latest news from Nahalal is that the decision has not yet been implemented. There has been no discussion in any forum of what functions each committee would perform or what the legal authority would be. While a decision to cease paying membership dues to the Moshav Movement has been taken, this does not in itself indicate that Nahalal has ceased to be a moshav. But then, the Nahalalim have never been able to agree on what they mean by moshav.

## NOTES

1. Second Aliyah. *Aliyah*—Immigration into Palestine and, later, Israel. Literally a "going up." Prestate immigration between the years 1882 and 1939 is divided into five periods known as Aliyot (plural of Aliyah). Thus: First Aliyah 1882–1903; Second Aliyah 1904–1914; Third Aliyah 1919–1923 (World War I intervened between the Second and Third Aliyot); Fourth Aliyah 1924–1931; and Fifth Aliyah 1932–1939. Each Aliyah differed from the preceding one, both in the immigrants' ideas and aims and in the conditions under which they came to Palestine. Nahalal's founders belonged mainly to the Second Aliyah, which was a wave of individual immigrants coming largely from middle and lower-middle class homes in Eastern Europe. Greatly influenced by the Russian socialist and revolutionary movements, they nevertheless saw the implementation of these ideas in a Zionist context.

2. *Kvutza:* a collective farming village, the prototype of all collective settlements, most of which today are known as kibbutzim (singular, kibbutz). The first kvutza, Degania, was founded by a small cohesive group of Second Aliyah agricultural laborers who began to cultivate a tract of land on the west bank of the Jordan in 1909. Nationally owned land, self-labor and management, and absolute quality of membership were its basic principles. The term has now come to denote a numerically small kibbutz.

3. Eliezer Joffe: the man generally acknowledged to be the ideological father of the moshav ovdim. In another culture, he would have been known as its prophet.

4. Among other considerations, the circle was seen as the most apt physical expression of a *Gemeinschaft* made up of equally based, family farm units and equidistant from the village center that housed the institutions and services. The service workers lived in an inner circle, surrounded by the farmers and their farmyards.

5. Naomi Nevo, *Social Change in a Veteran* Moshav Ovdim (Ann Arbor, Mich.: University Microfilms International, 1986). The historical depth that I have tried to give to present change has been taken from this study.

6. Nahalalim: members of Nahalal. Singular: Nahalali.

7. Moshavim: plural of moshav, cooperative small holders' village.

8. Quotas: In the mid-1950s, production quotas were introduced into the already highly planned economic system of Israel. Quotas were distributed

to each moshav and kibbutz (the agricultural commune). The moshav had to distribute the quotas to each individual farmer according to its own criteria. In Nahalal, the quotas were based on past achievements in the branch with an added category of starters. Without going into more detail here, it must be quite clear that this led to intense competition in Nahalal which was won by the strong, who thereby became stronger.

Chapter 7

# A Note on Even Yosef in the Early 1990s

*Harvey E. Goldberg*

*Harvey Goldberg's classic monograph (1972) on the community he calls Even Yosef examines the changes experienced by a group of Libyan Jewish immigrants to Israel. This chapter takes us back for a brief view of yet further changes since the 1960s. The trends remarked on here appear to conform with those found elsewhere in agriculture and village organization; but cultural continuities are also noticeable, as Goldberg, a uniquely qualified observer, points out.*

This sketch is based on two brief visits to Even Yosef (a pseudonym), a village first founded as a moshav by immigrants from Tripolitania, Libya, in 1951 (Goldberg 1972). These visits, of several hours each, were carried out in May 1990 and April 1992. They reflect a tour throughout the moshav and detailed conversations with one key informant.

Since completing my original field work in the village (in the spring of 1965), I have not conducted research about Even Yosef itself. I have visited the village from time to time, however, and have maintained relationships with the man cited as my key informant. These visits were part of a research effort to learn more about the past of the Jews of Tripolitania through oral history interviews

(Goldberg 1990). At the same time, very general information on developments in the village came to my attention in the course of these visits.

My key informant, by now a friend, was one of the first people I met in the village in late 1963. At that time, he had a job in the moshav office that included various kinds of recordkeeping, including the updating of a population register. I discovered that he had personal knowledge of almost all the names in the book (about 900 individuals). Later, I learned that this was a type of activity in which he had been engaged in the past and for which he had a personal inclination. The Allied forces' conquest of Libya in January 1943 enabled foreign Jewish organizations to work in the country on behalf of the local Jewish communities (De Felice 1985). My friend, then in his late teens, had worked in the local synagogue keeping records of allocations by the Joint Distribution Committee. His clerical position in the moshav, therefore, was not purely accidental. Even when he was no longer in that post, he always was interested in providing me with as precise and detailed information as possible.

I thus am confident that the information provided herein is reasonably accurate and can well serve the purpose of a general sketch. Many of my questions were asked twice—first in 1990 and then in 1992—and the answers were thoroughly consistent. When figures such as "about ten," or "20 to 30 percent" are cited, these are not based on calculations but are quotes from my friend's account to which I attach enough significance to commit them to print. I have not always added the terms "approximately" or "about," but all the figures reported should be understood in this sense.

One of the main characteristics of the moshav's population from the time of its founding was rapid population growth (Goldberg 1972). All of the ninety-one farmhouses of the moshav and all the resident or professional houses were inhabited, and the families had many children. This created pressure as early as the 1960s, and just before my arrival in the moshav, nineteen young families had moved to another moshav about three-quarters of an hour away by car. This trend has continued, with the younger generation moving to a number of new locales.

From 1963 to 1965, four young couples from Even Yosef were already living in a neighboring Yemenite moshav, becoming owners of farms there. This number now stands at six. Within the moshav, some of the original owners have passed away, and in other instances, an official transfer has been made to the inheriting son. In 1990, my friend estimated this as having taken place in

20 percent of the moshav farms, and in 1992 he gave the figure of 30 percent. Today also, all the houses (close to ten) that had once been available for residents (as opposed to farmers) have been turned over to younger families from the village. Just recently, a small housing project was completed near the local (state religious) school that serves Even Yosef and the two Yemenite moshavim in the area, and one Tripolitanian family has moved into this neighborhood. About ten years ago, six young families moved to a moshav in the Gaza Strip and continue to reside there. For the most part, however, younger families from Even Yosef have sought to remain close to their home moshav.

The most common pattern in this regard is that thirty to forty families from the moshav have taken up residence in the neighboring town, just over five minutes away by car. Some of them have bought apartments, and others rent. The men of most of these families (75% of them) continue to work in the moshav, while others work in jobs such as an agricultural worker, clerk, construction worker, cook, driver, or electrician. The term "families from the moshav" refers to families formed by sons of the moshav. About twenty daughters also live in the town, married to men who have no connection to Even Yosef. They have no work relationship to the moshav, nor are they candidates for inheritance. Another thirty daughters live in the nearest city, about a half hour away by car, and eight to ten sons live there as well.

I do not have information on the official population of the village today, but it most likely has declined in number. Many of the homes which were full with children in the mid-1960s are relatively empty. This is particularly noticeable as many of the families had added additional rooms to accommodate growing families, and often these rooms are no longer in daily use. On the other hand, the geographical proximity of the "expanded family" (Goldberg 1972) and the fact that many sons living in town work in the village probably are associated with frequent visiting by a floating, partially present population.

The village has not changed significantly in its physical appearance over the past twenty years. Many houses were enlarged as described earlier; and in the late 1960s, some differences began to appear in their outer appearance, such as the addition of decorative stucco. However, there are only a handful of cases in which a striking new structure, such as a house elevated on pillars, has been built. There has been a definite shift with regard to public buildings. The village *tzorkhaniah* (cooperative store) is no longer functioning, and most purchasing is done in the neighboring town (most of the families own a car or pick-up truck). A supply warehouse, formerly part of the cooperative, is now privately run. A youth clubhouse, once relatively neglected, has been enlarged and seems (based on a quick tour) to be one of the few public buildings looked after with some care (another is the clinic). An additional synagogue has been built, and synagogue attendance continues to be an important aspect of social life.

In the 1960s, the main agricultural branches in the village were citrus groves, vegetables, and poultry (Goldberg 1972). Since then, the cultivation of flowers was introduced. Gradually, sons became more active in running the farms. While formally still in charge of the farm, some of the older parents took paying jobs on the outside, often in unskilled or menial work.

My revisit in May 1990 took place about one year after the citrus groves had been uprooted, leaving bare the land that surrounded the *A plots* adjacent to the houses (Goldberg 1972). The A plots were covered with greenhouses for flowers. This continues to be the pattern. Agricultural activity is almost entirely devoted to the cultivation of flowers for export, accompanied by some raising of poultry (estimated at 5% of the economic activity). The A plots are almost completely utilized, usually by their owners. In 10 percent of the farms, this land is rented to someone else in the village (or a village son in the town).

The raising of flowers is labor-intensive, particularly during certain seasons. The farmers are dependent on the labor of Arab workers. This pattern was already in evidence in the early 1960s, when Arab women from neighboring villages were employed at vegetable-picking time. Today, many of the Arabs of that village are successful farmers, also participating in the export of flowers. Some of the Arab workers now employed in the moshav are from the area on both sides of the Green Line (the Israel–Jordan Armistice line). They reach the moshav through Arab labor bosses. At the beginning of each week, a bus-load of additional workers arrives in the moshav from the Gaza Strip and splits up among various farms. Some individual laborers have long-standing work relationships with some of the moshav members. In a number of instances, small huts have been built near the farm owners' homes, and the workers live in these structures during the week. Several of these huts carry graffiti in Arabic on their exterior walls, clearly pointing to their inhabitants and their function.

As in other moshavim, the cooperative nature of the village is now greatly attenuated and close to nonexistent. Credit for the initial investments of a growing season is no longer available through the purchasing organizations. Arrangements for packaging, shipping, and marketing abroad are made by each farm unit on its own. Each family is responsible for paying its own water bills, national insurance, and health insurance.

As elsewhere, members of the moshav are waiting to see the outcome of the debates and decisions concerning the arrangement to reduce moshav debts. When this issue became a problem, only about five or six farms were in the black. The rest were in debt. My friend mentioned NIS30,000 to 50,000 as typical of the amounts owed. His own debt is about NIS70,000. There are instances in the moshav of debts close to NIS200,000 and above. His criticism of the arrangement as it stood in April 1992 is that it only reduces the debt of those who owe a great deal but is of no help to those, like him, who owe a "medium amount."

There is still a general assembly of all the (male) moshav members, but it meets infrequently. Often, well over a year goes by between meetings. The general assembly elects the moshav committee, which (according to my friend) still exists but has no important functions. For a certain period of time, there was a committee appointed from the outside (by the Minstry of Interior?).

One of the striking features of the moshav in the 1960s was the centrality of the *mazkir hutz* (secretary for external matters), who retained the post for many years. He continued in this position until the mid- to late 1970s. After that, there were a number of outsiders appointed, but most remained in the job for only short periods of time. My friend sees one of these, who served in 1984, as particularly responsible for creating the situation in which members of the moshav signed for one another's debts.

A few years ago, one of the sons of the earlier mazkir was appointed to the position and seems to function satisfactorily. He does not deal with the matters of individual families (formerly closely linked to the cooperative organization) to the extent that his father did. This man also owns a large general store in the neighboring town in partnership with his brother. Like his father, he seems to combine the characteristics of both an insider and an outsider (Goldberg 1972).

Even Yosef has been buffeted by the travails throughout the country while still preserving some of the cultural features that characterized it thirty years ago. The maintenance of close geographical ties by most of the sons and daughters is a striking feature of that continuity, which is expressed in other ways as well. For example, two synagogues in the neighboring town are frequented mainly by the young families from the moshav.

In the mid-1960s, it was becoming standard for all youngsters in Even Yosef to finish grade school, and only a few continued to high school. Today, all youngsters have some postelementary school training. Children of the moshav still study in the rural state religious school, but many of those in the town also attend a religious elementary school. High school takes youngsters out of the village and the nearby town. Some have gone to religious schools, while others have attended secular institutions (I have no figures, but guess the latter are in the minority). As of 1990, no one from the village had entered an institution of higher learning.

Traditionalism is also evident with regard to the behavior of women. As of 1990, no young woman had served in *sherut le'umi* (volunteer national service for women not obligated to serve in the military because of their religious commmitments), to say nothing of army service per se. No women had learned to drive, and the question had never arisen as to a daughter becoming the owner of a farm. On the other hand, it seems that the older widows in the moshav have no interest in remarrying because, according to my friend, they have no financial problems. These patterns reflect an unselfconscious traditionalism, neither in the name of Libyan Jewish culture, nor linked to any explicit religious ideology. Over the past twenty years, I have been involved in many ethnic Libyan events in which Libyan Jews from the cities are active, only meeting a man from Even Yosef (the former mazkir) attending these events on one occassion. In the 1960s, the village was firmly linked to the Po'el Mizrahi party, with only a few deviants from this pattern, but today, according to reports, votes go in many different directions. The people of Even Yosef still retain the ability of remaining themselves while changing slowly and pragmatically in the heart of Israel's many economic, social, and cultural currents.

## REFERENCES

De Felice, R. 1985. *Jews in an Arab Land: Libya, 1835–1970.* J. Roumani, trans. Austin: University of Texas Press.

Goldberg, H. E. 1972. *Cave Dwellers and Citrus Growers: A Jewish Community in Libya and Israel.* Cambridge, Mass: Cambridge University Press.

———. 1990. *Jewish Life in Muslim Libya: Rivals and Relatives.* Chicago: University of Chicago Press.

# Chapter 8

# Generations Divorced: The Mutation of Familism among Atlas Mountains Immigrants in Israel

*Moshe Shokeid*

*Long-term research in a community of immigrants from the Atlas Mountains revealed the reluctance of farmers to allow their youth to share in the village resources or let their own children stay on with them in a father–son partnership. The apparent banishment of the young and the breakdown of the Moroccan pattern of household composition has been interpreted as a manifestation of the conflict between a cultural ethos and new environmental constraints. The chapter introduces some methodological queries and the ethnographic intricacies involved in community studies when deciphering processes of transformation, the reshaping of institutions, and the modification of values.*

My follow-up study since the 1960s in Romema,[1] a community transplanted from the Atlas Mountains to an Israeli Negev village (Shokeid 1971) has revealed unexpected paradoxes during field work and later at home when considering its interpretation and theoretical relevance. During the 1960s and 1970s, I was perplexed at observing the chaos in ritual, the discord among the worshipers, and the absence of the still revered traditional religious leaders from their native community in Morocco, who, although they had also arrived in Israel, had drifted away from the new Negev settlement into a more heterogeneous environment.[2]

In the mid-1970s, I recorded another observation—the subject of this chapter—the impact of which I grasped only a decade later. That more subtle and long-term process that affected intergenerational relations also carried immense consequences on the Romemites family and communal life. Although the villagers seemed to have preserved their patriarchal authority at home and in communal affairs and seemed also to have kept close affective ties with their offspring, nevertheless, they have persistently resisted the pressures exerted by both external agencies and their own maturing sons to let their youth stay on in the village. Expressing a wish to preserve farms for the younger children, they effectively expelled their older and, eventually, also their younger sons from their paternal home and from their natal community. Why were these God-fearing men and loving fathers now also banishing their own sons? Was I witnessing the Laius and Oedipus drama being enacted in the fields of the Negev?[3]

According to the Romemites' evidence, including both the former traders and draftsmen, sons were expected in Amran (as I named their community in Morocco) to stay on with their parents or with a few of their siblings. The firstborn son (who was also the offspring of his mother's first pregnancy) was entitled to two shares of his father's inheritance. All other male siblings received equal shares.[4] The distribution of a trader's wealth was a two-stage process. As the family patriarch aged and the younger generation expanded through marriage and procreation, the old man might decide on the first division of his possessions. That allocation contained mainly his movable trade assets, such as the bulk of his livestock, grain, nuts, and oil, as well as the land rented from Muslims. The father himself also took one share equal to the rest of his children (except for the firstborn, who received two shares). A rabbi might be consulted whenever a difficulty evolved during this division (such as the question as to the inheritance rights of grandsons).

With such relinquishing of goods, the father also gave up his leading position in managing the family business, but he still remained owner of considerable property (sufficient to ensure his economic security and his children's submission). Thus, after the first division, the Sebags' head of family of two generations ago kept on the buildings and land he owned in the nearby town of

Demnate. (Only after his death did his sons share the rest of his fortune.) By the time of the first division, the father's household had also been split. The Sebag trader mentioned above had nine sons. Two of his married sons had left their parents' household before the first division. But after that, five brothers between them established two independent joint households. The split of the father's household and the later split of siblings' joint households were caused by demographic growth, the discontent of wives, or the death of the father or his wife, who took charge of domestic affairs involving several daughters-in-law. (See Deshen 1989.) All brothers or only some of them, regardless of their domestic arrangements, might also continue to own jointly property allocated at the second division and compensate in-kind siblings who wish to take out their share. Household and inheritance arrangements were less elaborate among the poorer craftsmen's families. But here too, sons usually stayed on with their parents or siblings and cooperated in domestic and economic affairs.

In Israel, however, the former Amranites did not support the village youths' wishes to acquire farms and stay in Romema, nor did they succumb to the pressures exerted by the settlement authorities to share their farms with one child who would gradually take over the family farm and inherit it from his parents after their death. Therefore, Romema's children were actually eased out and forced to seek their livelihood and residence elsewhere.

I shall describe this phenomenon and try to explain its causes and effects. This case may reveal some dialectical linkages in processes of sociocultural transformations and reflect on our method of comprehending processes whereby sociocultures persist and change over time. My discussion also indicates the forces and rationale that instigate communal and individual greediness. In addition, it manifests the continuing merit of long-term community studies in recording and comprehending the processes that evoke the transformation and reshaping of institutions or patterns and the modification and adoption of values.

## BACKGROUND OF THE RESEARCH

I first studied Romema in 1961 when the Land Settlement Department engaged me as a rural sociologist to find ways of moderating the continuing internal strife in the village and to deal with the refusal of its settlers to take up farming. I returned to Romema in 1965 for an eighteen-month period of field work (October 1965 to April 1967) after the settlers had taken up farming. Ten years later, in 1976, I returned for a three-month summer stay. I last revisited Romema during 1987/88 in a series of intensive short visits. I met most of my acquaintances, went to see many of their children who had settled elsewhere, and collected detailed information about the first and second generation of Romema residents.

The initial physical and farming plans for the moshav (small holders' cooperative)[5] of Romema—founded and built in 1957—considered its resources sufficient for seventy-eight farms; the quota of water allocated to the village was based on that number of farms. Thus, each farm was entitled to forty dunams (8.2 acres) of irrigated land for seasonal planting and five dunams (approximately 1 acre) of irrigated orchard. In addition, the farmers were collectively entitled to the strip of land surrounding the village that is planned for dry farming. The orchard and the land for dry farming have been cultivated and managed by outside agencies and the profits equally divided among the farmers. Every moshav settler in the Negev is entitled to 21,000 cubic meters of water per year. At an early stage, Romema was entitled to the full quota of water for seventy-eight farms, regardless of the actual situation of farm allocation, which never surpassed a total of fifty-three.

In strong contrast to my observation of enterprising farming activity and prosperity of the mid-1960s and 1970s,[6] farming activity and economic expansion had declined dramatically since the mid-1980s. Many, if not the majority of moshavim, have in recent years gone through a severe economic crisis. Among the more obvious factors have been the collapse of the farmers' regional financial institutions[7] and the skyrocketing inflation and interest rates that left many farmers burdened with enormous debts. But these constraints[8] did not affect all moshavim equally. Thus, for example, the majority of Romema's farmers did not take advantage of the extended credit generously and indiscriminately available, particularly during the early 1980s, and which eventually ruined the economy of many villages. Therefore, only a small minority among the Romemites were burdened with crippling debts.

Most noticeable and, as I concluded, having the severest impact was the fact that the majority of the younger generation had left for other places. The continuing exodus of the younger generation, which had already started during the heyday of the moshav, has turned Romema into a *moshav zekenim*, an "old-age home," or a *ma'abarah*, a "shanty town"—terms often used by younger and older people in the village. Thus came to an end a golden age of about two decades of prosperity in Romema. Gador, the nearest settlement to Romema, a moshav about two kilometers away, was built in 1958 and populated mainly by veterans (either born in Israel or immigrants who had arrived in the country before 1948 mainly from Europe—often defined as Ashkenazim). Although Gador had been settled after Romema, its farm development and prosperity have been a permanent model for comparison among most Romemites. During my field work in 1976, in both villages, farms were considered property of high economic value, as expressed in similar phrases: *hayom meshek ze esek*—"today a farm is a business" (or "an intangible asset"). In Gador, however, all planned farms were occupied by active farmers. Moreover, the settlement authorities informed them they would not get an additional quota of water even if they decided to make additional farms available by giving up some of the land allotted to their own farms. Although there were as yet no young candidates for immediate settlement in Gador (whose settlers were all of the same young cohort when they founded Gador in 1958), they were nevertheless already considering the future texture and demography of their community. During my occasional visits to Gador in 1976, one of my acquaintances there whose children were still in school but among the oldest in their village cohort told me

> It seems inevitable that we shall share our farms with our children. True, with intensive farming two families can manage together economically, sharing a two-dunam greenhouse and an orchard on the rest of the farm plot. However, we would pretend that we could not accept this solution and shall continue to claim additional resources.

Not so in Romema, where only a few among the younger people were considering at that time the future composition of the community. "Within ten years the village will become an old parents home," declared the most enterprising among the few new farmers, who were allowed to share in the village resources and who wished to introduce changes in Romema's agriculture and modernize village life. (He thought, for example, that the farms allocated years ago to elderly parents should be reallocated to younger candidates.) But most other settlers were not worried by the inevitable departure of the young candidates. In most families there were younger children who could take over vacant farms one day or share their parents' farms when they became too tired to carry the responsibility alone.

## THE ALLOCATION AND INHERITANCE OF FARMS IN MOSHAV COMMUNITIES

The moshav framework as designed by its founders in the 1920s and enforced by its legal constitution allows for only one child to inherit the parents' farm.[9] That rule was intended to prevent the partition of farms into small units that would not be viable. This system of inheritance has worked quite successfully in most veteran villages (established before 1948 and mainly populated by immigrants of European extraction). The moshav supporting institutions offered plans and financial help to allow the inheriting sibling to stay (*haben hamamshikh*, "the one who stays on") by way of building an additional apartment for the young couple on the family farm. In these villages, parents and children have accommodated (though not always smoothly) to a pattern of sharing resources and the gradual decline of parental involvement in the running of the farm. Nevo's (1978) ethnography, which introduces detailed descriptions about the process of inheritance in a veteran moshav, clearly demonstrates the viability of that community in spite of the many difficulties involved. An inheritance process has evolved to overcome the rivalry of siblings and the potential tensions in a two-generation partnership. Moreover, a multifamily farm pattern emerged in cases of sibling partnerships.

During my stay in Romema in the summer of 1976, fifty-three farms had already been allocated. Ten of these had been taken over by new farmers two years earlier. The latter included two older heads of households who had refused to take up farming during the 1960s and a group of eight maturing sons of Romema farmers. These were the last farms allocated in Romema despite the initial planning and the resources available for seventy-eight farms and mounting pressure some years later by other maturing sons aspiring to acquire farms in Romema. That issue was among the major observations of my stay in 1976.

The Romemites' reluctance to add new farmers to their community was clearly identified by inside and outside observers alike. Thus, for example, during an interview I carried out in July 1976 with the official in charge of the Land Settlement Department's Negev office, he said:

> When you approach the farmers these days and try to discuss the problem of youth who have come of age and wish to settle on farms available in their own village, they reply that they must consider the future prospects of their four-year-old boys who might need a farm sixteen years from now. At the same time, they don't think at all about the son with whom they might share their own farm.

He was pinpointing the Romemites' (and other villagers of Middle Eastern and North African extraction) refusal to share with new farmers extra resources already made available to them although these were children of their own communities. Moreover, they were not yet planning for the day they would start sharing their own farms. The settlement institutions, however, no longer had the authority to force the villagers to implement the initial planning and allocate the resources under the moshav's control. The following observations, which were carried out in 1976, record the futile efforts of Romema's maturing sons to acquire farms in their home community.

## WAITING FOR FARMS IN THE 1970S—A BATTLE LOST

In the summer of 1976, six Romema boys were eager to take on farms. But the chairman of Romema's village committee[10]—himself the youngest among the founding generation—fully confirmed the observation made by the Land Settlement Department officer quoted earlier. Thus, he told me,

> There is no chance today for the new candidates; they must confront the strong opposition of farmers who want to preserve farms for their own young sons. I implemented the last allocation of farms in 1974, and I carried out that decision in spite of the resistance that was already there. But I can't repeat that fight all over again. I told my own brother he had no chance here.

No doubt that brother was disappointed, since another sibling in their family had been among the ten who received a farm in 1974. The chairman went on to tell me he had suggested to that younger brother that he should look for a farm in one of the new villages planned for young settlers in the nearby Gaza Strip or move away to the West Bank. He dismissed his brother's reluctance to part from his family and friends, assuming that he would find in the new villages the company of youth who might better suit his own education (he was then completing his studies in social work and was employed as a youth instructor).

Only one farmer, the father of only one son, had ever hinted at the possibility of sharing his farm with him one day. In 1976, there was already one farmer who had no child left in Romema because by the time one of his sons wished to remain, he could not get an additional farm. Another older farmer, the father of only one son, reacted indignantly at the suggestion made by representatives of the Land Settlement Department that he share his farm with him, saying, "Am I already dead?" As it turned out a few years later, the farmer who in 1976 seemed to be willing to share his farm with his son no longer encouraged his son (after he got married) to stay on with him and his mother.

Joseph Sebag, a school principal and the first Romemite to occupy a leading position in the region, received a farm in 1974 and was soon elected to the village committee. He repeated the same explanation that apparently caused the Romemites' reluctance to accept the aspirations of the new candidates: "The farmers think about their young children," referring to those farmers whose children were still too young to claim a farm. But he added another explanation: "During the last period of taking on new farmers, the candidates represented, more or less, the major factions [patronymic groups] in the village. But this time, the candidates do not form a similar combination of group representation." He thus implied that intravillage kin politics were also involved in the process of the allocation of farms. The inability of the candidates and their close relatives to introduce strong political pressure by way of a factional feud (which had dominated village life in earlier conflicts) made the farmers' reluctance easy to sustain.

A few days later during a meeting of the village committee, Joseph clashed on that matter with his own older brother, Daniel, father of several young children. Daniel argued that the issue should not be on the agenda of the committee but instead should first be discussed and decided upon by the General Assembly (a forum of all the farmers).[11] Joseph, himself only recently married, insisted that the committee must first decide and take its decision to the General Assembly for final approval or amendment. Otherwise, he claimed, the Assembly would certainly oppose the whole idea. But he remained in the minority. Bitterly, he told me later that he wanted the new younger members in order to prevent "the freezing of the moshav." He went on:

> My colleagues [on the committee] are shrewd. They pretend they have no personal complaints against any candidate: nevertheless, they decided to take the issue to the General Assembly. They know very well it has no chance there whatsoever. Even if only a few participants were to make some noise of disapproval, it would be a dead case.

The planners who at that time were considering the installation of greenhouses in Romema (which required the Settlement Department's financial support) recommended that project as a strategy to facilitate the sharing of farms by father–son families. The same planner who discussed this future development with me also criticized the Romemites' tendency to employ "too many Arab workers." He was implying the inexpensive labor arriving from Gaza since the Six Day War. Obviously critical and reflective, the planner nevertheless did not seem to notice the discrepancy between the goals of his planning and the labor situation on the farms. Even in advanced age, the Romemites were not obliged to resort to the labor of their children. As it actually turned out a few years later, during the 1980s, the greenhouses had no impact on the farms' household structure. In most cases, the farmer's wife became fully involved in the greenhouse activities and a full-time Arab worker was recruited as her major helper.

A few days after the meeting of the village committee mentioned above, two of the young candidates came to see me. They had no illusions about the dominant attitude among the committee members and the prospects for their request at the coming General Assembly. One among them, a teacher at the nearby high school, commented bitterly, "They don't want new moshav members, firstly, because they don't want to share the orchard with us [planted for seventy-eight farms] and other benefits earned collectively. Secondly, they want to preserve farms for their younger children." The speaker, who was in his early twenties, ended angrily, "Imagine, the child is not yet Bar-Mitzvah [thirteen years of age] and I can't get a farm!"

The reluctance of Romema's farmers to share the village resources with the young candidates has greatly affected the opportunities of their youth and the latter's reaction to that situation. The mode of behavior manifested by the older generation that projected "self-interest" and "greediness" concerning communal resources seems reminiscent of Banfield's (1958) characterization of "amoral familism" and Foster's (1961, 1967) interpretation of competition and conflict in peasant societies. No doubt, the circumstances of the Israeli moshav are very different from those of the poverty-stricken south Italian village studied by Banfield. The moshav's "pie"—its resources and technology—have never been absolutely limited and constant as prevalent in most peasant societies described by Foster. Nevertheless, the inevitable departure of Romema's youth has been perceived by insiders and outsiders alike to be mainly the result of the farmers' inability to consider not only the interests of the community at large but also the wishes and welfare of their own children.

## DEPARTING TO NEIGHBORING TOWNS—A GENERATION LOST

On my return to Romema in the autumn of 1987, eleven years after I had witnessed the futile efforts of Romema's maturing sons to acquire farms, there were no candidates looking forward to a farm or residence close to their family. This phenomenon struck me immediately upon my arrival as I first went to see Nathan, whose farm was just at the entrance to the village. He was talking to an urban-looking young man. As it turned out, he was the son of another prosperous Romema farmer. Now twenty-four and single, he was employed at various temporary jobs. To my fuzzy question, "What are the youth doing today?" he retorted immediately, "They are all leaving!" To support that assessment, he counted five among the young farmers of 1974 who had already left. Two of these had built their new homes a few kilometers away in a nonfarming community attached to Gador and inhabited by teachers and other professionals in the services of Romema's regional organization. Both have given up additional income and other benefits as moshav members. One of them was the committee member Joseph Sebag. Two others had left for the nearby town of Netivot. One of them had been extremely successful in farming and was reputed for his love of farm life. When he moved out to Netivot, he built an impressive ranch house there. There was no pressing economic cause for his departure, nor was it necessary to reside in Netivot in order to run the clothing shop he owned there. The same was true of his age-mate, who also left for Netivot and was employed as a driver. The fifth in that group was the young entrepreneur quoted above who moved to Beer Sheva, where I was told he had built "the most beautiful villa in town." Some time later I met those who had left and heard their own stories.

The narrative employed at that first encounter resembled later reports I got from young and older Romemites. Moreover, it also often was manifested in the style of Moroccan narrative accounts as revealed by Rosen (1984); "At no time does he [a Muslim youth] say what he or anyone else thinks or feels; the dialogue he recounts focuses instead on what men say, what they do, and what social consequences result." That style of speech referring to the behavior observed has often used a plural designation, thus adding authority to the observation the speaker was commenting about.

On the same day of my first visit in 1987, I also met Shlomo Sebag's two unmarried sons, who were in their late twenties. I had often visited their father during the earlier periods of my work in Romema and knew them well. They were now employed in various farming jobs on their father's farm and elsewhere as tractor drivers. When I repeated my question about the whereabouts of youth in their village, they answered almost simultaneously, "The young people leave; what are they going to do in that sort of place without company *(bli hevre)*?" Their brother-in-law, a resident of another Negev town, Ofakim, who was also on a visit, added to our discussion: "Young people can't do anything here. Even in the synagogue they have no say. The older people should have followed the example of their age-mates in Ofakim who let the young take charge of all communal affairs." One of Shlomo's sons assumed that when he married he would move to Netivot and the second son thought he would finally settle in Beer Sheva.

I visited Levy Biton, a skilled farmer, a shrewd politician, and a devoted and authoritative father of ten children. All four married sons in his family had left for other places in the Negev (Netivot, Sderot, Beer Sheva, and Neveh Dekalim). To my surprise at that exodus, he told me, "What could they do here, they can't make a living from farming!" This was an unexpected reply from a man who was obviously successful in his farming ventures and who raised a large family in considerable affluence. The response of his own son who also attended explained the discrepancy in Levy's argument. That son (twenty-nine years old), employed in a bank in Netivot, who arrived to collect a child he had left with his parents, told me sharply,

> Give me one good reason why young people should stay here. In spite of all recent difficulties, I believe that those who want to farm can make a good living from agriculture. But who wants to take a farm or stay in Romema without farming? Is there any suitable company here for young people [*yesh po hevra*]?

His mother added ironically, "Only the old stay on here."

Yehuda Mahluf, a successful farmer whose four married children (three sons and one daughter) had left the village, told me,

> You wait and see, no one among the rest of our children will stay here [he had two more single sons in their early twenties and a teenage girl], and I don't want them to stay around! This is not a suitable place for them. We have worked from dawn to dusk, but they are spoiled. They have done well in school and can settle elsewhere. We talk a lot on the phone, I have a car and they too have cars, so I see them a lot anyhow.

His wife added,

> How would they stay here without the company of friends [*ein lahem po hevra*]? If four or five other young families of close friends had stayed, they might have also considered staying in the village. It is different in towns where in spite of the many old people there are enough young people with whom to socialize.

With no shred of bitterness, she told me at last, "The moshav has been good for us, so you will still find us walking in the mud for another twenty years; it is just the same in other villages we visit occasionally to see our relatives." Yehuda and his wife's references to their hard work did not imply, however, that farming has been a degrading occupation. On the contrary, the Romemites were proud of their economic achievements, and farming in Israel has always been associated with the aspirations of Zionism and with social elitism.

Another forceful woman, in her mid-forties and a full partner to her husband's farming and who many thought should have been on the village committee had women in Romema been allowed to hold office in the moshav organization,[12] told me she had dared to participate in one General Assembly. There she tried to convince her audience that "we need the young sons to stay on in the village for our own sake and for the benefit of the moshav." But she was angrily shouted down with the advice "you give up your farm for the young." She concluded, "This is the way of the old people here." Her two younger brothers had left for the nearby towns and her own first maturing son had recently gone on a trip to the United States where he could stay with relatives.

I went to see those who had received farms in 1974 but had left a few years later. I visited Shaul Sebag in his store in Netivot that he managed together with his wife and that left the energetic man with the feeling of being only partly employed. He told me he had been the organizer of the group of new settlers of 1974; nevertheless, he could not stay in Romema "because there is no one to talk to there" (implying the older farmers who are in charge of all moshav affairs). Had the older Romemites allowed the young to take charge of the moshav's affairs, he told me, he would gladly sell his house in Netivot and return to his farm in Romema because "I love farming more than anything else and I wish I could return." Joseph Sebag, whose ambition to take part in national politics would have been better served had he stayed on the moshav, told me about the circumstances of his departure to the nearby rural center; "You can't consider only matters of economic gains and other personal benefits; you must also care about the social and spiritual conditions of your life [*hamazav hahevrati veharuhani*] as much as the circumstances of your children's education." He went on to describe his failed efforts to introduce some change in Romema. "But," he ended his story, "you can't do anything when the people you wish to help oppose you." His wife, also of Moroccan extraction, added, "I was shocked when I first visited Romema [born in a town in northern Israel]. I felt as if I had left summer for winter." With no efforts made to enable the younger and better educated young Romemites to stay on, the village maintained the image of an immigrant settlement, not much different from its appearance and lifestyle in the late 1950s and early 1960s.

Another among the five young settlers of 1974, now a wealthy businessman in Beer Sheva, reminded me of his suggestions at that time to revitalize the village by the transfer of farms from aging members to young candidates. "I warned them [the village committee]," he exclaimed, "the village would turn into an old parents' home, and I wasn't wrong. It happened just as I predicted." After he moved out to Beer Sheva, he maintained strong ties with his parents and brothers, three of whom have already been employed in his expanding business. He also supported another brother (a mechanic) to start his own shop in Netivot. I went to meet one of the embittered candidates of 1976 introduced earlier who was now the neighbor of Joseph Sebag at the nearby rural center. He told me that "since the young candidates of my generation were rejected, the drive of other young candidates was cut off at the roots and so it has remained to this day. It all happened this way because the farmers didn't wish to share their water with new settlers!"

The instructor appointed in 1987 by the settlement authorities to try to improve Romema's economic and social affairs assumed that with some support and encouragement a few farmers might gradually resume full farming activity. But he had no prescription for altering his own major observation: "The children have left and only the elderly have remained, and they gradually reduce their farming activity" (see Table 8.1). Himself a settler in nearby Gador, he told me about the continuing pressure in his own village to support the building of a new suburb for the younger generation. According to this plan, at least one son or daughter from each family would be able to stay on. This plan of residence within the village was unrelated to the question of farmers and their children sharing

**Table 8.1**
**The Distribution of Romema's Farmers* in 1987 by Year of Birth**

| | |
|---|---|
| 1900–1909 | 2 |
| 1910–1919 | 10 |
| 1920–1929 | 16 |
| 1930–1939 | 12 |
| 1940–1949 | 4 |
| 1950–1959 | 2 |
| 1960–1969 | 3 |
| Total | 49 |

*Including only farm owners who reside in the village. Four farmers have left but still remain nominal owners (two among them are from the 1940–1949 cohort and two from the 1950–1959 cohort).

the family farm. Gador farmers assumed their children would prefer independence and the close company of their age-mates. The plan, which suited both generations, seemed to promise the continuity and viability of their community.

We have discussed at length the overwhelming tendency among Romema young men to leave. But one is struck no less by the data demonstrating their tendency to stay on in the Negev region (see Table 8.2). Only 16.7 percent of the married men and 31.8 percent of the married women have left for the metropolitan area of Tel Aviv or other more remote places. However, the higher geographical mobility of women is related to a continuation of the old pattern of patrilocal residence. I was surprised to discover that 51.6 percent of the men and 54.6 percent of the women have upon their marriage taken up permanent residence in the Negev towns, among which Netivot (about 15 kilometers away) has been the major attraction (26.7% of the men and 25.8% of the women). Another 15 percent of the men have joined a few rural neighborhoods inhabited by the personnel of regional services and institutions. Five of these nine men have settled in a community only two kilometers away from Romema. A few women, however, have married into neighboring moshavim.

The youth, who could not wait indefinitely for a benevolent solution within the village, have in the meantime enjoyed the gradual development of the educational system that improved their skills and competitive position in the labor market. In addition, the gradual expansion of various regional services and the emergence of some industries in the nearby towns offered them new occupational alternatives outside the moshav. Unlike the phenomenon in many other countries and contrary to the predictions of Israeli researchers,[13] the departing youth did not abandon their community in order to seek their fortune faraway from home in the metropolitan centers of Israel. Rejected from their native village, they nevertheless sought residence in the nearby towns, some of which had for many years been considered inferior, economically and socially.[14] Most of Romema's young migrants are economically comfortable; they compose an important part of the leading sectors of their towns of residence.[15] Thus, while the moshav seems to have reached the point of no return, it has contributed to the expansion and development of the nearby semiurban communities.

## A PROCESS OF COMMUNAL TRANSFORMATION

Three major phenomena have been clearly observed: (1) The Romemites have firmly resisted the sharing of their economic assets with their children. (2) Their banished children have not departed to

**Table 8.2**
**The Distribution of Romema's Adult Children by Sex, Marital Status, and Residence (Percentages and Absolute Numbers)**

| | Romema | Another Rural Settlement in the Negev | Netivot | Beer-Sheba | Another Town in the Negev | Central Israel | Northern Israel | Abroad | Unknown | Total |
|---|---|---|---|---|---|---|---|---|---|---|
| Men | | | | | | | | | | |
| Single | 72.2<br>(39) | | | 9.2<br>(5) | 5.5<br>(3) | 9.2<br>(5) | | 1.9<br>(1) | 1.9<br>(1) | 99.9<br>(54) |
| Married | 16.6<br>(10) | 15.0<br>(9) | 26.7<br>(16) | 10.0<br>(6) | 14.9<br>(9) | 13.3<br>(8) | 1.7<br>(1) | 1.7<br>(1) | | 99.9<br>(60) |
| Divorced | | | 100<br>(1) | | | | | | | 100<br>(1) |
| Total | 42.6<br>(49) | 7.8<br>(9) | 13.9<br>(16) | 10.4<br>(12) | 10.3<br>(12) | 11.3<br>(13) | 0.9<br>(1) | 1.8<br>(2) | 0.9<br>(1) | 99.9<br>(115) |
| Women | | | | | | | | | | |
| Single | 72.2<br>(13) | | 5.5<br>(1) | | 11.1<br>(2) | 5.5<br>(1) | 5.5<br>(1) | | | 99.8<br>(18) |
| Married | 6.0<br>(4) | 7.5<br>(5) | 25.8<br>(17) | 12.1<br>(8) | 16.7<br>(11) | 21.2<br>(14) | 10.6<br>(7) | | | 99.9<br>(66) |
| Divorced | | | 100<br>(1) | | | | | | | 100<br>(1) |
| Total | 20.0<br>(17) | 5.8<br>(5) | 22.4<br>(19) | 9.4<br>(8) | 15.3<br>(13) | 17.6<br>(15) | 9.4<br>(8) | | | 99.9<br>(85) |

remote localities, a move that might have actually separated them from their parents and other relatives in Romema and its vicinity. (3) We recorded the emergence of a new rhetoric of sociality employed by both parents and children that emphasizes the solidarity and the shared interests of the society of age-mates.

Descendants of both the former better off traders and the poorer craftsmen were reluctant to give away the extra resources allocated to the village or share their farms with their own children, even at the price of remaining with no children at home. The few fathers who encouraged their children to stay on with them were among those who owned two farms as custodians of their own incapacitated elderly parents.[16] Their expressed wish to save farms for the younger children was never tested. The Romemites have never appeared to press their children to stay on with them. Whatever its sources, that claim appeared, therefore, as a sort of strategy that provided a rational and a morally justified excuse for confronting the pressures exerted by both older children of the moshav and representatives of the settlement authorities.

On my last visit to Romema in the summer of 1989, as I was discussing with Shlomo Sebag (my closest informant among the older generation of the Sebag family) the household arrangements and inheritance rules in Amran, a young relative, the eldest grandson of Amran's last head of community, also attended our session. Like most other married young men, he had left his parents' home and gone to live in a nearby rural center. I asked him why he had not stayed on in the village. He retorted passionately, "Our parents did not embrace us to them (*hem lo khibku otanu*)." He was using a metaphor that combined the notions of affection, seduction, and force. He went on to tell me that Gador's farmers have already submitted a plan to the various offices involved, requesting a permit to build 100 apartments on Gador's land in order to accommodate their youth.

We confront an entirely unexpected paradox. The sharing of communal economic resources and the family farms with children was not an embarrassing issue causing conflict and tension among farmers of Ashkenazi extraction who had not themselves been raised in extended families.[17] But farmers of Moroccan extraction who had been raised in extended families and in Morocco had submitted to their fathers' or elder brothers' authority nevertheless did not induce their youth to settle with them and did not accommodate them with a father–son farm partnership, which to many observers seemed an easy and "natural" arrangement. That reluctance did not imply the loss of a close family orientation or the open expression of tension and estrangement between successive generations. On the contrary, I have elsewhere described the amicable relationships between parents and children in Romema (Shokeid 1986). After marriage, most sons and daughters often spent the Sabbath and the holidays with their parents in Romema. The overwhelming preference of Romema children to reside in nearby localities in particular attests to the binding forces that still affected close relatives.

A most revealing observation has been the introduction of a new interpretation for the departure of youth whereby young and old were expressing a novel norm of sociability that competes with the paramount authority of family commitments. Although this new orientation did not actually transform the primary norm of family association, it nevertheless supported the cognitive depreciation of its compulsive force in committing social preferences. The Romemites seem to have adopted the Israeli rhetoric of cohesion[18] embodied in the solidarity of age groups.[19] This innovation was manifested in the references by old and young, men and women, to the lack of *hevra* or *hevre*—the company of friends (rather than the company of relatives)—as an incentive to leave the village and settle in the neighboring towns. This principle of association stands in conflict with other guides to ascriptive relationships, family and kinship in particular.

Katriel and Nesher (1986) traced the origin of *gibush* (literally "crystallization" meaning "cohesion")—another term from the linguistic repertoire frequently used in the rhetoric of Israeli school culture—as rooted in the communal utopia of socialist Zionism. The metaphors of *hevra* and *hevre* in particular have been an indispensable vehicle in the language of the new Israeli *communitas*[20] that replaced the greedy claims of the Jewish traditional family and the ranked structure of the *shtetl*. The terms *hevra* (literally "society") and *hevre* (meaning "one's particular group of friends") form elementary metaphors of sociality in mainstream Israeli culture. The term *hevre* often implies the joint company of men and women and thus further enhances its rejection of traditional categories for social interaction.

## STUDYING THE PROCESS OF SOCIOCULTURAL MUTATION

Anthropologists have invested much of their work in search of sociocultural continuity and change. Their approach has been criticized for many deficiencies. Thus, for example, Smith (1982) argued that they have been handicapped by the conceptual dichotomy and the empirical study of tradition versus change, which caused them to neglect the phenomenon itself, namely, the process of continuity. Another problem difficult to remedy has been anthropologists' tendencies to resort to the study of sociocultural continuity and change mainly as their subjects', or their own cognition identified a problematic area of human life experience as manifestly relevant to relationships, values, and beliefs of an earlier phase. We also develop a keen skill at detecting continuity versus change when external constraints clearly operate to support continuity or to exert pressures for discontinuity. In addition, the unit and the setting of observation of a specific phenomenon of change might also be broader and more complicated than that implied by the definition of a field work project.

I assume that many of these deficiencies are rooted in the nature of the ethnographic venture, which is usually limited in time. Not many anthropologists can afford a lifelong association with their field of research or the occasional meetings with their informants in order to complete some missing data or test a hypothesis. We seldom actually admit the obstacle: "Of course, we are rarely able to devote such single-minded attention to the study of a particular society or problem over time that we can watch the process of continuity [or change] at work" (Smith 1982). Moreover, the anthropologist's imagination is stimulated—though limited—by detecting direct linkages between phenomena synchronically or diachronically relevant in compact social contexts. Even those who are engaged in long-term research may not identify the linkages between processes in various spheres.

Ethnographic research has often revealed the problematic relationships between successive generations, particularly in farming societies. Arensberg and Kimball's (1948) study in Ireland represents a pertinent example. Waiting for the old man's decision to give up ownership and control over the family farm had become a major event in his children's life. The Irish case demonstrates the immense difficulties in the sharing and transfer of authority in a small family farm. The competition between father and siblings over the limited goods available resulted in a deep estrangement, particularly between father and sons. However, the Israeli moshav formal and informal inheritance arrangements seem to have resolved many of the difficulties reported from Ireland and other peasant societies.[21] The reluctance of the Romemites to share the village resources and their farms with their children cannot be interpreted as an outcome of the larger demographic composition of many Romema families. Even fathers of only one son were equally reluctant to share their farms, while parents and children in large Ashkenazi families have adopted the moshav rules of inheritance, or even evolved multifamily farms (Nevo 1978).

It seems the Romemites could not adjust to the consequences of the pressures to share resources with the younger generation raised in Israel that might have threatened their authority in family and communal affairs. To the extent that people carry on past experienced cultural messages imprinted in their store of cognitive and emotional "life guide discs," the former Amranites have confronted a difficult test in Israel. Their traditional code of family organization expressed in a patriarchical order and its relevant patterns of authority and ownership and their transfer preserved the dignity and security of the elderly traders and craftsmen. Moreover, Amran's venerated religious leaders manifested the ethos that supported the structure of family and communal authority. But the agents of social control and moral support for the new household and communal order, introduced by the Israeli settlement authorities and their ideological commitments, were completely alien and a poor substitute for the traditional communal leaders.

The Romemites who actually expelled their sons have thus discontinued their past family organization. The farmers watched their children leave but did not lose their loyalty and affection. (In the same way, although they had eased out their rabbis, nevertheless, they did not stop adoring them.[22]) This immense incongruity between tradition and practice has been divulged—and to an extent tolerated—through a manifestation of the mainstream Israeli rhetoric of sociality originating from the ethos of socialist Zionism, the pre-state youth movements, and paramilitary organizations. But the use of that terminology by the older generation in particular raised the notion of cultural masquerading. An alien vocabulary that recalls a world of experiences and symbols totally foreign has been employed to inform and signify the inevitability of the departure of the youth and the inevitable discontinuity of the extended family household arrangements. Nevertheless, family ties (manifested in celebrations, reunions, and mutual help), as much as a deep attachment to a diffused concept of *masoret beit abba*, "the tradition of the father's home" (usually connected with synagogue life and to religious details connected with family ceremonial life on the Sabbath and the festivals), have remained a major factor in the former Amranites' social life (Shokeid 1984). For the majority among the young, their company has mostly remained the *hevre* of relatives and their close associates.

The research carried out among North African Jews in Israel has often revealed the continuing commitment to major social patterns and cultural symbols, in spite of their transformation in everyday life, or even their apparent abandonment.[23] Thus, for example, Deshen (1974), who observed changes in religious behavior,[24] interpreted them as an expression of pervasive feelings of religious failure and self-deprecation that buttress piety and religious status rather than the simplistic assumption of secularization. Have we not thereby gained a partial answer to the ethnographic riddle posed earlier? The Ashkenazi settlers who had not been raised under the matrix of a patriarchal order and its moral forces seem to have adapted more easily to a father-son partnership and at an early stage. Is it not the latent continuity and survival, in spite of a display of the loss and betrayal of the spirit of a patriarchal order (as manifested in the apparent greediness of the older generation),[25] that might offer an insight into the Romemites' reluctance to accommodate their children's desire and the settlement authorities' pressure? The former Amranites did not become an obsolete generation in Israel; they remained masters of their homes, their farms, and community. They preserved their paternal dignity, but not without cost: They have seen their children moving away—as their rabbis did two decades earlier—into a more heterogeneous environment. Divorced, but not estranged, separate but close, generous hosts and affectionate guests, the older Amranites and the younger Romemites have introduced their own script for a new-old predicament—the conflict of generations.

Community studies seem to have lost much of their attraction if not their legitimacy in the era of postmodernist and often ethnography-free anthropology. But as long as anthropologists choose to study communities in vivo, they cannot ignore Arensberg and Kimball's (1968) conclusions about the raison d'être of the method and theory of community studies. The significant questions are still those associated with transformation, the reshaping of institutions or patterns, and the modification of values. Arensberg and Kimball made an important contribution in perceiving community within the context of system analysis "as process based on the activities of individuals in events rather than as a thing of set structure and pattern" (1968). I would like to use the case introduced in this chapter to advance our perception of community as process based on the choices, decisions, and options taken by individuals, as much as by their ongoing discourse with prevalent ideologies and values, but which are rarely observed as events. More than ever before the raison d'être and the future of community studies relies on the ethnographer's comprehension of "his or her community" as highly intermeshed in space, in time, and ideas through dynamic relationships within a continuously expanding and changing environment. Unless we adopt a method to identify and interlock eventual and uneventual processes cutting across past and present communal borders and to unfold imprinted layers of ethos and sentiment, event analysis and other forms of system analysis in community settings may well resemble a fire alarm set in an airless room.

## NOTES

This ongoing research was facilitated by a grant from the Pinhas Sapir Center for Development at Tel Aviv University, a Ford Foundation grant through the Israel Foundation Trustees, and a fellowship at Manchester University financed by the Bernstein Israeli Research Trust. I am grateful to Andrea Tyree for the insightful contribution to an earlier version and to Shlomo Deshen for his valuable comments on that version.

1. All names of persons and places are fictitious.

2. I interpreted the departure of the old religious leadership as partly affected by the bureaucratization of religious leadership forced by external agencies but also no less as the result of intracommunal processes that have shattered the traditional social order of ranked patronymic groups that also displayed interconnecting ties between the religious and the politicoeconomic elites (see Shokeid 1979).

3. See, for example, Sheleff's (1981) thesis and review of literature about adults' hostility toward youth.

4. Daughters did not normally share with their brothers in the major divisions of their father's property. Women's position at inheritance was similar to that observed in Moroccan Muslim society (see Maher 1974; Geertz 1979).

5. The moshav, introduced in Israel in the 1920s by young European pioneers, is established on national land and organized on the basis of moderate economic-social cooperation. Economic cooperation involves, for example, the marketing of agricultural products, the supply of farm and household needs, and the sharing of dairies and granaries. Administration is by committees and offices elected by the village General Assembly. For a more complete description of the moshav, see Weingrod 1966; Willner 1969; Shokeid 1971; Goldberg 1972.

6. See Shokeid 1980.

7. These regional institutions have been *Irgunei ha'keniyot*, the "purchasing organizations"; see Shokeid 1980.

8. The integration of the occupied West Bank and Gaza into the Israeli economy since 1967 must have been another important factor influencing moshav farming and economy. Moreover, the change of government in Israel since 1977 that brought the Likud to power has also transformed economic policies and national priorities. The moshav and kibbutz movements were far closer to the policymakers and to national ideological priorities during the long reign of the preceding labor governments.

9. See Otolengy 1982.

10. The major executive body of the *moshav* organization (see Shokeid 1971).

11. See Shokeid 1971.

12. For more details about the position of women in Romema, see Shokeid 1971.

13. Don, Bar-El, Ilan, Bar Hai, Greenspan, and Schnitt (1977) predicted the inevitable departure of the younger generation in many moshav communities because of the lack of farms, in spite of its preference to remain. They assumed the better educated and better skilled in their cohort would be the first to leave and start a process of emigration. The researchers were convinced, however, that the exodus of youth would carry most children of the moshav away to the major metropolitan areas and, therefore, also augment grave social problems there (see also Ginsberg 1970).

14. The planning and development of immigrant settlement in the Negev has reached an unexpected cycle of events. A massive investment of economic and social resources, guided and supervised by the personnel of the Jewish Agency Land Settlement Department, as well as by the nationwide moshav settlement movements linked to political parties, was directed to support the moshavim. In contrast, the nearby immigrant towns were left to the support and changing policies of the various government departments responsible for different services (such as the Ministry of Housing, the Ministry of Industry and Commerce, or the Ministry of Welfare). See Deshen 1970; Spilerman and Habib 1976; Cooper 1978.

15. Compared with their age-mates born in these same towns, the young Romemites, children of a prosperous moshav, have for many years enjoyed better conditions at home. They could always get some work on their parents' farm or with other farmers, earn some money to spend or save, and look for new opportunities before they married. Many could also rely on their parents' financial support or their guarantees for a bank loan to cover particular expenses.

16. Their parents, who had no other sons, were already too old to manage their farms at the time of farm allocation.

17. During recent centuries, European Jewish society was undergoing a continuing process of sociocultural change which undermined the authority of tradition in various spheres of life (see Katz 1961).

18. See Katriel and Nesher 1986.

19. See Eisenstadt 1956.

20. See, for example, the repeated use of these terms in Ben-Yehuda's (1981) novel about her life as a participant in the Palmach (an underground organization during the British mandate).

21. See Foster 1961; Wolf 1966.

22. The absence of their Moroccan rabbis, however, was compensated by an expansion of the Romemites' repertoire of pilgrimages to visit living Moroccan sages and the tombs of scholars and saints from ancient and more recent times (Shokeid 1974b, 1979).

23. See Deshen 1974; Shokeid 1974a, 1979, 1984.

24. Such as the abandonment of the rite of *nefilat apayim*, the wearing of *talit katan*, or the growing of a beard (Deshen 1974).

25. Eickelman (1989) argued that my earlier account (Shokeid 1971) of conflict among the Romemites, which I thought was indicative of a breakdown of their prior community, has actually demonstrated a marked continuity of Moroccan values and assumptions concerning interpersonal obligations. I assume that my present account of dialectical responses to environmental constraints may represent a similar phenomenon in terms of Moroccan culture.

## REFERENCES

Arensberg, C. M., and T. S. Kimball. 1948. *Family and Community in Ireland*. Cambridge, Mass.: Harvard University Press.

———. 1968. "Community Studies: Retrospect and Prospect." *American Journal of Sociology* 73: 691–705.

Banfield, E. C. 1958. *The Moral Basis of a Backward Society*. Glencoe, Ill.: Free Press.

Ben-Yehuda, N. 1981. *1948—Between Calendars*. Jerusalem: Keter (Hebrew).

Cooper, S. 1978. "Newgate: An Old-New Town in the Negev." Ph.D. diss. Washington, D.C.: Catholic University of America.

Deshen, S. 1970. *Immigrant Voters in Israel: Parties and Congregations in a Local Election Campaign*. Manchester: Manchester University Press.

———. 1974. "The Varieties of Abandonment of Religious Symbols." In *The Predicament of Homecoming: Cultural and Social Life of North African Immigrants in Israel*, S. Deshen and M. Shokeid, eds. Ithaca: Cornell University Press.

———. 1989. *The Melah Society*. Chicago: University of Chicago Press.

Don, Y., R. Bar-El, B. Ilan, A. Bar Hai, I. Greenspan, and D. Schnitt. 1977. *Industry in the Moshav*. Rehovot: Settlement Studies Center (Hebrew).

Eickelman, D. 1989. *The Middle East: An Anthropological Approach*. Englewoods Cliffs, N.J.: Prentice-Hall.

Eisenstadt, S. N. 1956. *From Generation to Generation: Age Groups and Social Structure*. London: Routledge.

Foster, G. M. 1961. "Impersonal Relations in Peasant Society." *Human Organization* 19: 174–178.

———. 1967. "Peasant Society and the Image of Limited Good." In *Peasant Society*, J. M. Potter, M. N. Diaz, and G. M. Foster, eds. Boston: Little Brown.

Geertz, H. 1979. "The Meaning of Family Ties." In *Meaning and Order in Moroccan Society*, C. Geertz, H. Geertz, and L. Rosen, eds. New York: Cambridge University Press.

Ginsberg, Y. 1970. *Rural–Urban Migration in Israel*. Rehovot: Settlement Studies Center (Hebrew).

Goldberg, H. D. 1972. *Cave Dwellers and Citrus Growers: A Jewish Community in Libya and Israel*. Cambridge, Mass.: Cambridge University Press.

Katriel, T., and P. Nesher. 1986. "*Gibush:* The Rhetoric of Cohesion in Israeli School Culture." *Comparative Education Review* 30: 216–231.

Katz, J. 1961. *Tradition and Crisis.* New York: Free Press.

Maher, V. 1974. *Women and Property in Morocco.* New York: Cambridge University Press.

Nevo, N. 1978. "Social Change in a Veteran Moshav Ovdim." Ph.D. diss. Tel Aviv: Tel Aviv University.

Otolengy, S. 1982. "The Rights of Farm Inheritance in the Moshav." Working paper 14–82. Tel Aviv University: The Pinhas Sapir Center for Development Studies (Hebrew).

Rosen, L. 1984. *Bargaining for Reality: The Construction of Social Relations in a Muslim Community.* Chicago: University of Chicago Press.

Sheleff, L. 1981. *Generations Apart: Adult Hostility to Youth.* New York: McGraw Hill.

Shokeid, M. 1971. *The Dual Heritage: Immigrants from the Atlas Mountains in an Israeli Village.* Manchester: Manchester University Press (1985 augmented edition, New Brunswick, N.J.: Transaction Books).

———. 1974a. "The Evolution of Kinship Ties among Moroccan Immigrants." In *The Predicament of Homecoming*, S. Deshen and M. Shokeid, eds. Ithaca: Cornell University Press.

———. 1974b. "An Anthropological Perspective on Ascetic Behavior and Religious Change." In *The Predicament of Homecoming,* S. Deshen and M. Shokeid, eds. Ithaca: Cornell University Press.

———. 1979. "The Decline of Personal Endowment of Atlas Mountains Religious Leaders in Israel." *Anthropological Quarterly* 52: 186–197.

———. 1980. "Reconciling with Bureaucracy: Middle Eastern Immigrants' *Moshav* in Transition." *Economic Development and Cultural Change* 29: 157–205.

———. 1984. "Cultural Ethnicity in Israel: The Case of Middle Eastern Jews Religiosity." *Association for Jewish Studies Review* 9: 249–272.

———. 1986. "The Impact of Migration on the Moroccan Jewish Family in Israel." In *The Jewish Family: Myth and Reality*, S. M. Cohen and P. E. Hyman, eds. New York and London: Holmes and Meier.

Smith, E. 1982. "The Process of Sociocultural Continuity." *Current Anthropology* 23: 127–142.

Spilerman, S., and J. Habib. 1976. "Development Towns in Israel: The Role of Community in Creating Ethnic Disparities in Labor Force Characteristics." *American Journal of Sociology* 81: 781–812.

Weingrod, A. 1966. *Reluctant Pioneers: Village Development in Israel.* Ithaca: Cornell University Press.

Willner, D. 1969. *Nation-Building and Community in Israel.* Princeton: Princeton University Press.

Wolf, E. R. 1966. *Peasants.* Englewood Cliffs, N.J.: Prentice-Hall.

## Chapter 9

# From Village to Suburb: From Cooperation to Individualism

*Leonard Mars*

*Social anthropologist Leonard Mars returned to the moshav he had studied in 1967 to find that the one-time agricultural community had been transformed to a commuter's suburb. The dramatic event that contributed most to the change in the character of the community was the Six Day War. This ended the village's isolation as a border community, which now had easy access to Jerusalem and Tel Aviv. Now, there is division within the community between the minority whose interest continues to be in farming and those whose interest is restricted to principal services. A decline of support for cooperative ventures in agriculture may be related to other changes in social interaction.*

When I first entered Givah at the end of January 1967, the village was at the end of a cul-de-sac, since it was on the Jordanian border separated from Latrun in Jordan by three kilometers of no man's land. The old Tel Aviv–Jerusalem road of the Mandate period cut through the center of the moshav, but no traffic passed along it. The Six Day War dramatically altered Givah's geographical position, which was now in the center of the country, equidistant on the east–west axis between the two metropoles, with good communication along the north–south axis.

At that time, the village was inhabited by forty-seven families, although it had been planned to accommodate eighty. Many houses were empty, some were dilapidated, others used to store hay or to house Arab workers, and one eventually emptied of its hay to accommodate the anthropologist and his wife. Those houses that were inhabited by villagers were modest dwellings built by the Land Settlement Department of the Jewish Agency, but villagers were reluctant to invest in them or even to establish gardens around them. To explain this state of affairs, they told me that they "were sitting on their suitcases." They were, of course, quite right to be apprehensive about their future, especially since many of them were survivors of the death camps who had originated from Eastern and Central Europe, Czechoslovakia, and Romania in particular but also from a variety of other European, North African, and Asian countries. The village itself had experienced murderous infiltration from across the border that had resulted in the death of several villagers in the early 1950s and the subsequent exodus of some of the earlier settlers. In short, many villagers were not committed to spending their retirement in Givah, and some had in fact invested in properties in the coastal towns of Israel.

Givah, as was the case with nearly all moshavim in the 1960s, was based on agriculture, primarily on dairy farming, with branches of poultry, vegetables, and, in one case, a sizable flock of sheep. Some members combined part-time agriculture with outside work, while others derived their livelihood totally from urban employment or, in a few cases, provided services to the moshav's farming community. Politically, the village was dominated by the largest economic interest group, the dairy farmers, who were supported by the Land Settlement Department, which was responsible for settling Jewish immigrants on the land and represented in Givah in the person of Moses, whose role was multiplex, since he was a founder-member of Givah, though no longer a resident there, and also a representative of MAPAM, the political party with which the moshav was affiliated (Mars 1980).

When I returned to the moshav in 1990,[1] several changes were apparent. First, there were no vacant houses, since the village was fully settled. Second, there was increased variety in the housing stock—luxurious villas were now located next to the original, moshav, square-box houses. Third, industrial and service enterprises had developed. Fourth, the number of farmers had declined, especially in dairy production. Fifth, the village lacked an elected administrative committee and was administered by a committee appointed by the Registrar for Cooperative Societies (Rasham). Finally, cooperative village organization had collapsed. How are these changes related to each other and how can they be explained? Initially, the explanation is to be sought in demographic and geographic factors and concomitant socioeconomic forces.

## POPULATION GROWTH

Before the Six Day War when Givah was on the border, there had been a turnover of population; several families had departed from the village, leaving many farms and houses vacant. The village was reluctant to accept newcomers since those who remained, especially the farmers, had seized and distributed the production quota, water quota, land, and even the vacant houses of those who had left (Mars 1980). This situation indicates that local interests could and did thwart national goals of settling immigrants on the land. It also demonstrates that immigrants who had been directed to the village resisted the plans to settle them against their will in rural areas and moved out when an opportunity arose. Toward the end of my field work in 1968, the first five sons of immigrant farmers became members in their own right (Mars 1980) and two persons who had lived for some time in Givah as residents (toshavim) were also granted membership (Mars 1980). These new members were already insiders, especially the sons (indeed, one of the residents was also a son of a member). However, the five sons, who were only sons, were granted separate farms in defiance of moshav law that decrees they should wait until they inherit their parents' farms.

In the 1970s and 1980s, more and more children of the immigrant founders became members and were either allocated farms or, in some cases, auxiliary farms (meshek ezer).[2] A few outsiders bought farms from villagers who had sold and moved out, but most of the increase in population is due to the retention of the founders' children. Most of these children and newcomers did not farm, but worked outside the village in nearby towns and cities. In many cases, these children were professionals with a university education. They had improved their original homes and built luxurious houses surrounded by fences that enclosed gardens and guard dogs.

A moshav is a complex socioeconomic institution serving many purposes. Essentially, however, these diverse purposes can be reduced to two. First, the moshav is legally constituted as an agricultural cooperative, an aguda; that is to say, it is a farming community whose members own and operate family farms. Second, the moshav is a municipal unit that forms the lowest level of local government within the jurisdiction of the regional council (moetzet aizori). Within the moshav, the municipal functions are discharged by the locally elected administrative committee (va'ad hanhala).[3] As an aguda, the moshav represents and serves the interests of its members, many of whom may have renounced farming as a full-time occupation, but who nevertheless remain members of the village and who are enfranchised to vote in the elections for the va'ad and also entitled to stand for election to the va'ad. Not all villagers are members of Givah as an agricultural cooperative since there exist many toshavim of the village who require and seek municipal services but who are disenfranchised and hence not entitled to vote in elections to the va'ad. Their interests in the village as a municipal entity are strong; they pay local taxes that are set by the va'ad, but they are constitutionally impotent. Apart from these residents, there are other local people whose interest in the agricultural aspect of the moshav is slight. For example, there are members who have ceased to farm, such as pensioners, or those who have opted out of agriculture and commute to paid jobs in nearby towns, or those who work in the village in areas other than agriculture.

What has happened in terms of the social structure of moshavim is nothing new. Indeed, Yonina Talmon-Garber examined increasing social differentiation and stratification in cooperative villages forty years ago (1952), and I focused on this issue in my book. What is new, however, is the overall reduction in the number of farmers; the increase in the number of residents; the emergence of a category, possibly a class, of pensioners; and the development of industrial enterprises. In short, social differentiation has proceeded apace so that the range of interests has become increasingly diverse while the formal structure of the moshav has remained intact. The gap between the ossified constitutional and political structure of the village and the current socioeconomic situation is manifest in the inability of the moshav to elect a va'ad that had the confidence of the whole village and has resulted in the decision by the Rasham to appoint a va'ad (va'ad memuneh) to administer the affairs of the village.

The va'ad was originally predicated on the assumption that its members had been democratically elected by an association united by a common occupation that shared social values such as equality, mutual aid, and mutual liability. Although different interest groups were emerging in the 1960s in the village, the agricultural faction was still dominant and controlled the va'ad. The moshav had been officially designated in the 1950s as a dairy farming village and the Land Settlement Department had invested in the development of this branch of agriculture. Moses, the village administrator, attended their interests even though the tenets of the moshav movement were regularly infringed. The dairy farmers, who had been in office for several years, had maintained their control despite a short-term takeover in 1968 by a loose coalition that had held office when the dairy farmers and their supporters had complacently failed to vote in the annual elections. Elections to the administrative committee are held simultaneously with elections to the inspection committee (va'adat biqoret). This body consists of three members whose role is to scrutinize the members of the va'ad and to monitor their observance of the rules of the moshav movement. Traditionally, the inspection committee, though formally required by the constitution of the moshav movement, was elected in a halfhearted manner since village members were supposed to trust their fellows voted onto the va'ad by the General Assembly. Candidates were prevailed on rather than volunteered to stand for this committee and usually polled far less votes than those on the main committee (Mars 1980). In fact, the inspection committee was never intended to function in the classic moshav ovdim (Baldwin 1972) nor in the immigrant moshav (Shokeid 1971). One villager opined that the inspection committee was an afterthought and that only failures and those lacking self-respect were willing to stand for a position that lacked respect.

In Givah in 1989, the va'ad that still represented farming interests had come under increased scrutiny from an inspection committee that had taken its brief seriously. The inspection committee consisted of three people, two newcomers and one veteran, Benjamin, who had been a sheep farmer in 1967 but who was now the owner of a dairy—an industrial enterprise, that produced cheeses and yogurts for a small, upscale clientele. In the 1950s and 1960s, Benjamin had been chairman of the va'ad and was an able and skillful village politician. The two newcomers on the inspection committee were professional people, Israelis with long experience of Israeli society and its institutions. One had been a senior manager of the bus cooperative, Egged, a major national institution with a multimillion dollar annual budget. He had taken early retirement to Givah where his wife's brother had facilitated his entry. A systematic, well-organized man whose wife had died a few years after their arrival, he had commenced to probe the workings of the va'ad, publishing various papers and reports on his word processor and distributing them to the villagers. The third member of the inspection committee was an accountant, a woman whose husband, a part-time farmer, had been a member of the va'ad but who had been ousted by the chairman on the grounds that he had failed to attend three successive meetings and hence had forfeited his position.

These three constituted a well-qualified and formidable inspection committee which had made some awkward discoveries. One of them was a decision by the va'ad to raise the salary of the village secretary (mazkir) without the approval of the General Assembly. Effectively, this amounted to the passing of a budget unauthorized by the General Assembly. The inspection committee demanded that the budget be subject to public scrutiny and approval. Having failed to gain satisfaction, the inspection committee wrote to the Rasham who sent a representative to attend a General Assembly convened by the va'ad with the result that the Rasham, influenced both by the letter from the inspection committee and the treatment of his representative, decided to appoint a va'ad consisting of three outsiders who were to administer the village.

What is of interest is that though the locally elected va'ad had now ceased to function, the inspection committee remained in office. In local terms, the influence of the inspection committee had increased, as had that of the secretary. Members of the inspection committee and the secretary became the conduits through which villagers raised their concerns and cases with the appointed va'ad whose nonresident appointees only attended Givah for their weekly meeting and who knew little if anything about the histories and reputations of villagers and consequently their social and political credit.

The main political issue that divides the village and that resulted in the appointment of a va'ad by the Rasham is the conflict between agricultural members on the one hand and nonagricultural members allied with residents on the other. This conflict is about local taxation and the salaried personnel required to service the ever-diminishing agricultural sector. In effect, all villagers pay the same local taxes whether they are farmers or not, and these taxes pay the salaries of the village secretary, accountant and bookkeeper, and other village employees such as cleaners, vineyard manager, and security guards.

Members of the inspection committee represent a large section of the village who have renounced the cooperative ethos and who employ their own accountants, bookkeepers, and advisors to manage their business and personal affairs. Among these are the industrialists, the professionals, and the pensioners who have ceased to farm. They are of the opinion that local taxes could be cut if farmers employed their own professional consultants or if the administrative staff employed by the village became self-employed, working on a contractual basis for each farmer rather than employed by the aguda. Thus, the decision of the last elected va'ad to raise the salary of the mazkir without bringing the matter to the

General Assembly was construed as an example of partiality, favorable to the interests of the farmers and detrimental to those of the nonfarmers.

Support for the nonfarmers' case had come from the Rasham who had recommended that both the village bookkeeper and the mazkir should be switched from salaried employment with the Givah to contractual employment with individuals. Interestingly, the Rasham's own appointment, the head of the committee, while favoring the decision on the bookkeeper, had recommended the retention of the mazkir on a part-time basis that amounted to 50 percent of a full salary. The negotiations on these changes in employment were taking place in the spring of 1990 when I was in the field, so I do not know the final outcome.

Effectively, economic changes in the village had proceeded ahead of administrative change. The local situation was only a microcosm of the national scene that had seen agriculture decline as a major economic and ideological force in Israeli society. Again, the accountant on the inspection committee indicated the problem when she observed, "The organization of agriculture has created an inflated bureaucracy, such as the va'ad, the mazkirut, the bookkeeper, and so on." A more pithy formulation of that conclusion was put to me by Gideon Kressel who remarked, "You don't keep a cow in order to get a glass of milk."

## THE HOUSING STOCK

Since 1967, many of the moshavniks that I knew and, in particular, their children have improved their properties, and many newcomers have bought and renovated the original houses. In 1990 on my first full day in Givah, I rose early and took a nostalgic walk along the road where I used to live. I thought it would be a walk down memory lane. As I strolled along, parts of it seemed familiar and parts were strange so that I experienced feelings of dislocation and disorientation. The house that I used to live in was a flat-roofed, three-room, grey concrete box with a modest garden that I had begun to create. It had no fence and no lawn. Indeed, the boundaries of my property were so vague that a neighbor's horse at the end of its tether would graze under my window. That house no longer existed and on its site stood a magnificent, gleaming white villa surrounded by a high fence and patrolled by a fierce-looking but docile Alsatian dog. It had a beautiful lawn and behind the back of the house stretched an orchard of seven dunams. Proceeding along the road, I encountered a few more fine villas likewise surrounded by fences and guarded by dogs; but also there were several dwellings that were unchanged, hence giving me some sense of recollection, while the novelty induced senses of wonder and estrangement.

## PRIVATIZATION AND INDIVIDUALISM

I have mentioned the fences that surround the increasingly diverse housing stock of the village. The housing stock in 1967 was mainly of a uniform type since most of the original Arab dwellings had been demolished in the early 1950s and had been replaced by single-story square concrete boxes built by the villagers with funds from the Land Settlement Department. In the 1960s, villagers had moderately improved them by the addition of a sloping, red-tiled roof, funded by the Land Settlement Department, and by the addition of another smaller concrete room, thus giving an L-shaped form to the house. One or two families still lived in Arab houses that for some reason had escaped demolition, but the degree of differentiation in housing was negligible. By 1990, some of the red-tiled dwellings remained, but these were dwarfed and literally put in the shade by large, in some cases, two-story dwellings called villas, which were designed to the specifications of their owners, each slightly different from the next.

The housing stock furnishes an immediate and stark pointer to the extent of social and economic differentiation within the village. Economic differentiation was marked in 1967/68, but it was not so visible in terms of patterns of consumption (housing, cars, and telephones) because the manifestations of economic differentiation were not public. Even now in 1990, many of the veteran farmers still live in the neat, modest, red-roofed dwellings, while their children bask in greater luxury.

## POSTAL DISTRIBUTION

Whereas the housing stock presents an obvious index to the changes in the lives of villagers, there are more subtle and unobtrusive indications. For example, the system of postal distribution has changed. On the first day of my return, I had ventured briefly into the offices of the village secretariat, which were housed in an old Arab building situated in the center of the moshav. The main door was open and led into a small hall off which were the offices of the village secretary and bookkeeper, a layout identical to that of 1967. However, there was a small but significant difference—lining the walls of the entrance were rows of individual mailboxes, each bearing the name of a family from the village. In order to illustrate the importance of this development, I shall describe the system of postal distribution as it was in 1967 and as I observed it on the second day of my visit in 1990.

During my original field work, a mobile post van would arrive at the village at approximately 10:00 A.M. and hand—or even throw—a bundle of letters to the aged village postmaster seated under a tree in the center of the village surrounded by a throng of eager villagers, both male and female, who would gather there from about 9:45 A.M. onward and would often stay to chat and to gossip for some time after the distribution of the mail (Mars 1980).

A considerable percentage of the village would convene there, and if a letter was not collected, then either a neighbor would deliver it or the addressee would call in person to collect it later from the postmaster's home. The distribution of mail in the late 1960s was both public and collective. The postal tree was the major forum for collective assembly where gossip could be exchanged, information transacted, and villagers from different neighborhoods of the moshav reunited, not necessarily in a harmonious manner. Often it was the scene of teasing and joking relationships that represented the simultaneous expression of antagonism and solidarity.[4] That part of the day was a liminal[5] period when normal work ceased, a period of timeout that could vary from twenty to fifty minutes.

In 1990, the situation was completely different. The venue for distribution had shifted slightly, only about ten meters, but away from the tree to a new concrete bus shelter—a switch from nature to culture. The postmaster was no longer the old widow (since she had died) but a younger, married woman who, I was told, did not always appear for the distribution; and indeed on the day I observed the scene, she turned up after the time the mail was scheduled to appear. The composition of the group had changed markedly. The small groups that assembled were almost all veteran Hungarian settlers, mainly women, but there was a handful of men (e.g., a retired farmer who accompanied his wife and a man who in 1967 was then a confirmed bachelor employed as a nurse and now married). The one working farmer who appeared was Dan, who operated his farm with his son, and he too appeared only at the end of the distribution. Not all of the veteran Hungarians gathered, but the core of the group came from their ranks and represented a form of continuity with the past. One of the older women still collected the mail for some of her veteran neighbors. "You can't trust those letter boxes, children play there, remove letters, and fool around," she explained to me. At first I was skeptical about this allegation because I suspected she was disgruntled, but later, others (mainly newcomers) complained to me about tampering with the mail and damage to their boxes, which I had noticed.

The post that was scheduled to arrive at 10:00 A.M. finally arrived at 10:30 A.M., though some of us had been waiting since 9:40 A.M. After the post's arrival, the postmaster took the letters and inserted them in the individual boxes in the village office. What these boxes signify is the increasing privatization and individualism already seen in the style of housing, coupled with fences and guard dogs. The location of these boxes in the heart of the village's administrative offices tells us something about that administration and its clients. Given that so many residents work outside the village, it is not surprising that the postal distribution has changed. The presence of the boxes together in one central point represents a practical solution to a problem and also signifies minimal neighborly ties. However, there is one newcomer who for his business has taken out a mailbox in the nearby town of Ramla rather than receive correspondence in the village, to which his private mail is addressed. The separation of residence from work that this indicates had already occurred when villagers began seeking outside employment, but they still chose to have their mail sent to the village.

## THE DECLINE OF COOPERATION

I have documented the rise of individualism and privatization among inhabitants of Givah, nevertheless, as an aguda, the moshav retains a corporate, legal identity and still exhibits some cooperative features, which I shall examine shortly. What remains of cooperative farming in Givah today?

Two weeks of intensive reading on moshavim that I had undertaken in Sde Boqer prior to visiting Givah made me aware of the decline of agriculture in general, and of cooperative agriculture in particular, since my original research. As with the letter boxes, there were physical but much more public manifestations, namely, in the two buildings that in 1967/68 had provided cooperative services to the farmers of the village. First, the village dairy (Mars 1982) that had received the farmers' milk and where I had been briefly employed was in a dilapidated state. Its equipment had been removed and its walls breached; in short, it was an eyesore serving no purpose except as a memorial to times past. Second, the village warehouse, from where farmers used to buy their feed and seeds supplied on a cooperative basis by the Purchasing Organization (*Irgun Keniyot*), no longer operated. The building still existed, but it had been converted into a social club named after a young soldier from the village who had been killed in the Lebanese War of 1982 and whom I remember as a bright-eyed, intelligent, six-year-old in kindergarten.

The demise of the one and the transformation of the other clearly signaled the end of the village as an economic cooperative. However, the village as an aguda has two enterprises that it owns and that are run by the village as a whole. The first is an orange grove near Ramat Gan, thirty kilometers from Givah, that is leased to a holding company. Originally, this grove was allocated to the village by the Land Settlement Department, which recognized that Givah itself was not suitable for growing citrus fruit and which sought to guarantee an income to an immigrant moshav from what was in the late 1960s and early 1970s a profitable branch of agriculture. The second venture, also funded by the Land Settlement Department in the 1980s, was a vineyard on village land that in 1968 had been utilized by a small, cotton growing consortium (Mars 1980). These two enterprises are owned by the village as an aguda because of the difficulties in parceling out the land among its individual members and because a national institution, the Land Settlement Department, allocates budgets to the village as a whole. As the accountant on the inspection committee explained to me, "The *mazkir* is strong because it receives funds from institutions. If budgets were personal or individual, then the power of the *va'ad* as representative of the *aguda* would decline."

Not only does the aguda receive funds, but it also receives production quotas and a water quota. For example, the Givah milk quota that in 1968 was distributed among fourteen farmers, is now in the hands of three. These three have increased their productivity and have formed their own subcooperative, The Milk Producers of the Valley. As noted, they have dispensed with the central milk depot and each has his own refrigerated milk storage plant; is responsible for his own supply of feed; and individually arranges veterinary services. The national milk marketing company comes to each farm to collect its output rather than each farmer transporting his milk to the central depot to await collection. The three farmers distribute the production quota among themselves, have their own bank account, and in effect constitute a cooperative within a cooperative. Two of them are veteran villagers in their late sixties and the third is a son of a veteran who has taken over his father's dairy farm. Within the next few years, it is highly likely that the youngest of the trio will acquire the production quota of the whole village, though constitutionally the aguda, effectively the va'ad, retains the power to withhold part of the quota and allocate it to another applicant.

Other manifestations of individualism and privacy are to be found in the field of communications, especially in the acquisition of private telephones, cars, computers, and television sets. In 1967, the whole village was limited to a single phone located in the village office. From 1968, individual villagers began to acquire their own personal phones. The first was Benjamin who recalled in 1990 the inconvenience he suffered from neighbors who wished to use it. By 1990, the village had its own internal telephone directory that listed subscribers. These phones indicate an increased contact with the outside world and also change in the nature of communication within the village. Villagers are in touch with family and relatives abroad both through correspondence and overseas vacations and also via the telephone. The diminution in face-to-face contact at the village level, seen in the small congregation at the postal distribution, has enabled more impersonal communication to take place by phone. At the same time, however, the phone may have encouraged villagers who wish to call each other. When I stayed in Benjamin's house and I wished to accept the invitation extended earlier in the day by a neighbor who lived opposite, Benjamin's wife suggested I telephone to tell them I would be coming. This was to advise them that I would be entering their territory and moreover to protect me from their fierce, as I thought, Alsatian dog.

Communications for villagers have been improved by the acquisition of private cars, which were scarce in 1968. What motorized transport existed was for agricultural purposes (e.g., tractors, vans, and a lorry owned by an individual who would deliver produce for villagers). The village administrator possessed a car and so did the chairman of the va'ad, but nobody else had a private car.

Many villagers, especially the young professionals, have their own computers, whereas the village office lacks one; here again we see the private sector outstripping the public.

## SUMMARY AND CONCLUSION

My book based on field work in 1967/68 documented the conflict between various village factions based on farming and nonfarming interests. At that time, the dairy farmers, the largest single interest group, controlled the committee with the support of the village administrator, who was employed by the Land Settlement Department. Even then, the hegemony of that group was coming under challenge. On my return in 1990, I found a much altered village. Not only had the number of dairy farmers declined, but the way in which they organized their branch had changed. Cooperative services at the village level to meet the needs of the dairy farmers were nonexistent; the milk depot was derelict; the warehouse transformed into a clubhouse. The three dairy farmers had formed themselves into a subcooperative to which the aguda had allocated the village's milk quota. Payment to these dairy farmers no longer went through the village office but entered straight into the bank account of the subcooperative.

It seems likely not only in Givah but in other moshavim that the trend will be away from a single agricultural cooperative to either a number of loosely structured subcooperatives or to individual production only.

The particular decline in the number of dairy farmers was accompanied by a general reduction in the number of farmers, twelve in total, compared with nine industrialists, twenty pensioners, and thirty-seven outside workers. Many of the outside workers engaged in agricultural endeavors on a part-time basis, operated in branches compatible with their jobs, such as fruit orchards.[6] What is particularly significant is the reluctance of the founders' sons to practice agriculture. Only two sons have followed their fathers' occupation, and, revealingly, neither of them had proceeded to higher education, unlike their contemporaries. Givah confirms Zussman's (1988) conclusion that the moshav has retained educated and skilled members but that these same people have opted out of full-time farming. Shokeid's data on a moshav in the Negev, however, shows a different pattern. There, most of the village youth had migrated to the nearby development town of Netivot (Shokeid 1989). The rather banal conclusion to be drawn is that what happens in a particular moshav depends both on internal and external factors and indicates the diversity among settlements even when subject to common national factors.

The development of a highly educated, independent-minded village population, whether recruited from outside or from among the children of the founders of the village, resuscitated the dormant inspection committee that came to voice the interests and values of the nonfarming members and of the residents. These interests center on the moshav as a municipal institution rather than as an aguda catering to an association of farmers. Future development would suggest that the present aguda will be dissolved so that farmers will either form their own subcooperatives, such as the dairy producers, or that agriculture will be organized on an individual basis in a few hands after the agricultural land has been redistributed. Land may well be taken from the present aguda for the building of houses to attract newcomers to the village. This is a move favored by the young professionals. The village could then be reconstituted as a social cooperative (*aguda kehillatit*) or simply become a municipal village, which would enhance the movement already made from a village to a suburb or the greater Tel Aviv conurbation.

## NOTES

1. This return was made possible by an invitation from Professor Gideon Kressel of the Jacob Blaustein Research Institute of the University of Negev, Beer Sheva, Sde Boqer campus. I am grateful for his support and also for assistance from Dr. Moshe Schwartz in reviving my interest in moshavim.
2. Meshek ezer: These were houses with small plots of land set aside for employees resident in the village who were not entitled to membership.
3. Hereafter referred to, following village practice, as the va'ad.
4. See A. R. Radcliffe-Brown (1952).
5. See V. W. Turner (1967).
6. N. Haruvi and Y. Kislev (1989) point out that the expansion of fruit orchards between 1971 and 1981 coincided with the decline in the numbers of dairy farmers.

## REFERENCES

Baldwin, E. 1972. *Differentiation and Co-Operation in an Israeli Veteran Moshav*. Manchester: Manchester University Press.

Haruvi, N., and Y. Kislev. 1989. "Structure and Specialization in the Family Farm in the Moshav." Working paper No. 25. Rehovot: Development Study Center.

Mars, L. 1980. *The Village and the State*. Westmead: Gower.

Radcliffe-Brown, A. R. 1952. "On Joking Relationships." In *Structure and Function in Primitive Society*, pp. 90–104. Glencoe, Ill.: Free Press.

Schwartz, M., R. Bar-El, R. Finkel, and A. Nesher. 1986. "A Moshav-Based Industry in Israel." Working paper No. 16. Rehovot: Settlement Studies Center.

Shokeid, M. 1971. *The Dual Heritage: Immigrants from the Atlas Mountains in an Israeli Village*. Manchester: Manchester University Press.

———. 1989. "The Decline of an Immigrant Moshav: The Exodus of the Young." Discussion paper 2–89. Tel Aviv: Pinhas Sapir Center for Development.

Talmon-Garber, Y. 1952. "Social Differentiation in Cooperative Communities." *British Journal of Sociology* 2: 339–351.

Turner, V. W. 1967. "Betwixt and Between: The Liminal Period in *Rites de Passage*." In *Forest of Symbols*, pp. 93–111. Ithaca: Cornell University Press.

Zussman, P. 1988. *Individual Behavior and Social Choice in a Cooperative Settlement*. Jerusalem: Magnes Press.

Chapter 10

# "He Who Stays in Agriculture Is Not a 'Freier' "—The Spirit of Competition among Members of the Moshav Is Eroded When Unskilled Arab Labor Enters the Scene

*Gideon M. Kressel*

*The significance of work, especially manual labor in agriculture, is of central importance to the ideology of the moshav. It is this issue that Gideon Kressel explores as it has changed through time in a community he has studied for decades. What has been the social impact of the employment of Arab wage laborers to do agricultural work in the moshav? What are the differences in perception between older and younger generations?*

## DILEMMA AT WORK

One works for a living, but the function of work in higher forms that concerns us here can be derived from two basic aspects under which we meet it: as a contest for something or a representation of something. These aspects can be linked in such a way that the work represents a contest or else becomes a contest for the best representation of something. Representation means display, and this may simply consist in the exhibition of something naturally or culturally given before an audience.

Competition at work in Zionist-Socialist Israel was for the establishment of a strong and just society of people living by their own labor (i.e., turning the Jewish occupational pyramid upside down[1] or implementation of Hebrew labor) to consolidate the new nation-building. Jewish immigrants to Eretz Israel were competing at work in pioneering, deliberateness, dedication, perseverance, risking of their own resources, and in putting their life in danger. They competed at work with body strength, determination, dexterity, proficiency, knowledge, and sacrifice.

Usually, the contest is not devoid of purpose; that is, the action does not begin and end in itself: It contributes to the necessary life-process of the group. Moral values then stress physical and intellectual fitness as part of the work ethic. The more apt it is to raise the intensity of life in the individual or the group, the more readily it will become part of civilization. The object for which we compete is first and foremost victory, and victory is associated with the ways in which it can be enjoyed. The fruits of victory may be honor and prestige, or they may be material or symbolic values that affect the life of the worker and the welfare of the whole group.

Competition over outstanding achievement in Israel in the 1990s does not affect the workplaces as it has in the past for two reasons: (1) the spirit of capitalism in its local expression removed the stress from the way you make your fortune and placed it on the amount that you amass. (2) This process was accompanied by the increase of Arab labor at the workplaces alongside the Jews. The Arabs' entrance, at first to infrastructure jobs, diminished the willingness of members to compete at work even where they could do better than Arabs. The view of the workplace as a racetrack suggests an addition to current sociological explanations of the choice of future occupation and of vocational behavior.

Cooperation is requested for the maintenance of public institutions and in extending a helping hand to members in need.[2] Competition has been confined to implementing skillful, diligent and productive agriculture. A change of values in the Israeli society and, recently, in government policy pertaining to the rural sector has disrupted these two tracks. Infusion of individualism and capitalism into the arteries of the moshav has brought about the dismantling of cooperative institutions and the liquidation of mutual guarantee agreements. Mechanization through the 1960s increased work efficiency, and the saturation of markets, quotas on production imposed by the Ministry of Agriculture, and the law of inheritance (which prevents fragmentation of the small holding) spurred competition in which the better fit were supposed to survive. However, in moshav Noon Yud (pseudonym), an unexpected change of circumstances has encouraged the tendency of the most able entrepreneurs to quit farming. Employment of Gazan workers in moshav farms beginning in the summer of 1968 has disrupted the conventional code of fairness in competitive behavior based on self-labor.

While the innovation of modern agricultural technology distinguished the most capable farmer and enhanced the attractiveness of farming, the farmers' reliance on unskilled labor harmed their professional reputation. Now, ambitious members who strive to demonstrate excellence must seek to achieve this outside the moshav.

Since striving for excellence is strongly inculcated, many people in their search for arenas of performance leave farming because they do not find it challenging enough. *Homo competitivus* will endeavor to accumulate capital, scrupulously implement socialism, or strictly obey religious commandments, finding it difficult to live without criteria for achievement. Anomie is explained in this book by confusion in the "rules of the game" and a lapse of controls ensuring fair play whereby cheating prevails (Huizinga 1950).

By 1970, 400 moshavim, 250 kibbutzim, and a variety of private farms were supplying the local market with large quantities of farm products, and since opportunities for export remained limited, the number of people that could be profitably employed in agriculture was proportionately restricted (Gvati 1981). To maintain sufficient income, there was increased effort in mechanization to save on manpower and to increase work efficiency. Family members whose work was no longer required looked for employment elsewhere; they joined the armed forces or the civil service. The laws of inheritance, which prevented the splitting up of the small holdings, also encouraged the tendency to leave as a way of providing a variety of sources of family income. State institutions provided loans on easy terms to improve homes and public buildings, thus offering some compensation for the widening gap between rural and urban incomes. The road system was also improved to shorten the time to and from the towns.

Then, the income from external work outgrew that from the small family farm. The improved quality of village life thus became suited to the raising of families; young couples began to acquire houses from farmers whose own children had grown up and had preferred the comforts of the city. The average age of the population of the moshavim went down, and initially, along the coastal plain, the proportion of residents (those owning no farmland or with no on-the-spot income) gradually overtook that of the farmers who remained (Applebaum 1974, 1979, 1983; Shoresh 1988). A few efficient farmers took over the cultivation of plots belonging to those employed outside, exploiting technological advances and hiring farm workers (Zussman 1988). In this way, the principles of equal distribution of productive capacity were broken; the community, which had measured diligence and manual dexterity in terms of material success when everybody occupied farms of similar size, lost its yardstick for comparison. As the balance tilted toward the nonfarmers, there was also a change of values; Arab labor became more acceptable.

## NOON YUD, FROM MOSHAV TO KFAR TO SUBURB

The first decade of statehood was marked by a great flow of immigration, by massive investment in establishing new moshavim, and by the encouragement of movement from town to the country in order to increase agricultural production and to overcome the austerity situation and need for rationing. But by the end of the 1950s, agricultural production had grown to the degree that the Ministry of Agriculture had to apply strict production quotas to small holdings;[3] it also favored the moshavim on the periphery over those in the center of the country. In Noon Yud, a moshav of the "Ha-Ihud Ha-Hakla'i," in the center of the country founded in 1954 on an area of 2,400 dunams (400 acres), the family farms were dismantled one by one.

At the beginning of the 1970s, three dairy farms were still operating, and by the end of the decade, all had closed down. Of some sixty-five poultry houses in the 1950s, only five were still used in 1991. Of forty-five families that earned their living from farming, three are left. Of some 200 families in the village (inadvertently, the term *kfar* [village] was found in use to replace the term moshav), thirty-six still gain some income from agriculture—from the cooperative orchard (560 dunams) and from family orchards. Overall, the number of full-time farmers totals ten; these include vegetable growers (whose fields cover the entire arable area of the village, plus additional plots belonging to other settlements, on which watermelons are grown), poultry farmers, a young man specializing in greenhouses, and a young man growing mushrooms.

Noon Yud's founders, immigrants from Argentina, understood the importance of economic diversification and set aside an area for crafts and workshops, allocating small plots to members who did not intend to devote themselves entirely to agriculture (maximum area of six dunams, compared to complete small holdings of twenty-eight dunams). When most of the original members abandoned agriculture, whoever did not take the opportunity of developing a small craft or industry turned to trade or employment outside the farm. Houses changed hands as well. In these circumstances, the standard of living of those working outside rapidly improved in comparison to those who kept on with farming. This trend continued, sometimes with the letting of prime plots lie fallow or neglect of farm buildings and installations built at great expense. Keeping open the option of directing unsuccessful farmers to more suitable branches is not easy to accomplish; professional training and capital are needed to develop alternative branches in highly advanced agriculture, while fluctuations in the cost of credit (as in 1984 and 1985) are liable to overwhelm the borrowers and threaten their chances of success.

Most of the investment in developing the industrial area therefore came from outsiders, some of whom even settled in the village. In this way, local residents could find alternative employment close to home. The settlement of people of means added to the village's coffers, improved its status, and attracted more new settlers of means. The demand for houses, even those that had remained unimproved since originally being handed over to the settlers as Jewish Agency houses in the early 1950s, pushed up their value. Whoever had previously wanted to leave but had not had the means of doing so, including those in debt whose farms and homes had been mortgaged, were from now on able to find buyers and thus set themselves up in a town.

The recession in the Israeli economy of 1965 to 1967 and the resulting unemployment strengthened the members' desire to hold onto their homes and even returned people to their fallow fields, but not for long.

The Six Day War extricated the economy from the recession, improved the employment situation, and at the same time transferred emphasis in agricultural development to new settlements beyond the Green Line. In the veteran moshavim, agriculture as a percentage of all types of employment was reduced (Applebaum 1986), which prompted their further suburbanization. The moshavim in the central area close to arterial roads and the cities were in particular demand, although not all to the same extent.

According to Applebaum (1986), with members, the demand for housing was particularly great when (1) the cooperative was strong enough to prevent members from becoming embroiled in debts (because property under mortgage does not attract buyers until the owners' debts are met); (2) they were not too strong to the point of enforcing mutual guarantees among the members for payment of each other's debts, something that keeps people of financial means at arms length; (3) questions about equality among members of independent work or of Hebrew labor were suppressed and did not overly concern the moshav institutions; (4) the community did not place a premium on its agricultural character and actually limited the spread of livestock farms by (the smell of) which townspeople tended to be put off; (5) a well-defined community character was maintained, the moshav chose not to absorb more than a few families at a time, did not permit private building to encroach on public areas or fields, prevented multistoried buildings, and so on; (6) the community was zealous in maintaining living standards and keeping up public institutions and surrounding greenery and controlled the cleanliness of public areas; (7) out of responsibility to the local authority, the growth of competitive power centers was discouraged and the development of special interest groups and struggles between Hamulot (groupings of related extended families) was prevented.

In Noon Yud, all the aforementioned elements were generally present, it was highly in favor of housing, and was the most prosperous of the villages in the area. Its outstanding characteristics were the following:

1. It exhibited a stance independent of party politics or movement ideology. Despite its linkage to the Ihud Hakla'i (a national organization of middle-class settlements founded in 1944), it was not obligated to the organization, which in any case was weak and made fewer demands of its settlements, than, for instance, the Moshav Movement. Even though most of its members were formerly in pioneer youth movements in their countries of origin and included former kibbutzniks who did not conceal their inclination toward the values of the labor movement, their demands of the moshav institutions were limited to the subject of public services. The moshav secretary, not a resident, was hired by public tender, and an interview to assess his or her qualifications for the job according to his or her education and managerial skills was conducted.
2. It supplied "green lungs" and cheap seasonal fruits to the residents in the extensive orchard areas (mostly citrus [Grossman 1987]). The limited number of livestock farms (that nevertheless supplied neighbors with low-cost produce) and the industrial factories set up according to a master plan at some distance from the residential area offered additional income and products at bargain prices to the residents.
3. It was secular but liberal; that is, supportive of those of the members who followed religious traditions and ceremonies but noncoercive. The village had a synagogue and an occasional *minyan* (a minimum of ten

adult men, a quorum necessary for community prayer), mainly on religious festivals and, in particular, on the high holy days; it was also widely used for family ceremonies such as bar mitzvahs and memorials.

4. It was Ashkenazi in character; while dozens of members were from families that came from Arab countries, or themselves emigrated from an Arab country, they mostly displayed outstanding economic success and a Western approach to planning and family and to their children's education. In comparison to the moshavim of the oriental aliyah (wave of immigration), the members of Noon Yud resisted the tempting offers of grants and subsidies by the settlement authorities and thus did not allow them to run their businesses. They had to achieve their goals by their own efforts. They also kept strict control over the cleanliness of public areas, prevented the development of internal feuds to the point of undermining the authority of the cooperative, did not allow members to be late in paying debts, and sometimes even forced those who did to sell their farms and leave the village.

## ALTERNATION IN THE APPROACH TO AGRICULTURE

The founders of the village remember years of toil and trial and initial hardship. They saw it as one of the obligations of the individual to bring about the needed statistical change in the national character (an indirect relationship to Borochov's model of "inverting the pyramid").[4] "This was required by the reality of the situation if one wanted national independence." They saw it as a way of fighting against the negative stereotype of Jews in the Diaspora and therefore as a means of changing the image of the Jew held by the non-Jew. It was a *mitzva* (a good deed) essential for the renewal of the people's connection with its soil, to prove its right to its land, and so on.

For the sons of the founders (i.e., the generation that grew up or entered the circle of employment in Noon Yud in the 1970s), manual labor was viewed as nonessential. Of fourteen high school graduates who were asked in 1972 about their plans for the future after their army service, only two indicated that they looked to agriculture or an academic profession connected to it. The others felt that agriculture was only fitted for the young while they are preparing themselves for a serious, "real" profession. In one's twenties, one must learn a profession to replace agriculture. "A person who respects himself," they said, "cannot remain in agriculture beyond the age of thirty. When one is established one ought to be far, far away from it."

A son of the moshav, a student in the Haifa Technion, described what passed through his mind until he settled on this point of view based on events that influenced him when he was positioned to do farm work in a kibbutz in the Jezreel Valley along with his military service:

One winter's afternoon, I was shepherding the Kibbutz herd of sheep on the hills to the south, and heavy rain began to fall. At first I thought that it was stopping; I sat behind the stone terrace wrapped up in my poncho, till I noticed that everything was soaking, from my poncho to my clothes, to my boots, till I was like a sopping rag, and it went on and on and on. When you're on guard duty in uniform and with a weapon, and that happens to you, at least you know you are on duty, "standing under fire and rain." But I was dressed in blue work clothes and it made me think of my father tending his bit of orchard in Noon Yud. I felt as if he was, at that moment, in this flood, soaked to his bones and wretchedly dragging himself home. I pitied him so much, and myself put in his position—when I'd already be old, with children a respected individual—and looking like that. That's when I decided that farming was not for me.

The devaluation of the status of manual labor and of farming, which then provided employment and a source of income for many of the members, came as a surprising disappointment, which was received with deep sorrow. With the pain came the realization that the farmers must analyze and understand what was happening to answer the question, "Why are the young not following in our footsteps?" Their answers are varied and can be divided into those blaming themselves and those blaming outside factors in general and the leaders of the country in particular. Below, one of the founding members reflects on the situation and tries to explain why the second generation is uninterested in the farm. Ya'acov, a fifty-five-year-old production engineer with Tadiran, explains (recorded in the autumn of 1978 while orange picking in his orchard):

Keeping a full orchard farm [twenty-eight dunams] today, with technical equipment and with growing and organizing arrangements [using contractors who employ wage laborers], is less of an effort than it used to be when we started . . . , production costs are less than they were. All the same, if you want to save money and you've got the energy, you can always go out and do it right yourself . . . That way the harvest is complete [the workers always leave fruit on the trees], the pruning is more responsible [one doesn't break branches like the laborers], irrigation is more economical [if you see a leak, you fix it at once], and the things you do with responsibility and love are felt at once. It also pays in the long run, perhaps not much. That is why [to make up for the low income relative to the effort] the traditional subsidy was granted [that the Labor Alignment Governments paid to the farmers]. So now, when they are reducing that subsidy, how can they complain about the youngsters? The present government is harming the profitability of farming, and the understandable reaction is pullout. Where great effort is optional, show me anybody who chooses the hard way. Look at yourself, for example!

When I finish work at the factory I'm happy to change my clothes and come out here till sundown . . . for me it's like sport and recreation, looking after my orchard although it brings in no income. I'm even ready to spend the money I earn at Tadiran on this "hobby," on the understanding that it goes in cycles and that next year we'll succeed, and what we earn will pay for what we lost this year . . . but if we have to keep absorbing losses year after year, my wife, Esther, won't agree.

According to Ya'acov, doing things without getting a reasonable profit from them hurts the pride of youngsters who can't overcome the values of Israeli society in general and can't allow himself or herself to be presented as a *freier* (sucker, an idiot—a term of considerable contempt in modern Israel).[5] For himself, again, fruit-growing is not viewed as employment in the sense of work to earn a living but an occupation, a way of preserving his health and sanity. Michael (sixty-five years old) also belongs to the original group of settlers, a former kibbutznik who works for the Histadrut (the General Federation of Labor) controller. Like Ya'acov, this "afternoon farmer" places the blame on the country's leaders. However, he puts particular emphasis on American values pervading Israeli society as being responsible for the results, not only the poor income from farming. He says that while it is true that he himself is not dependent on the few pennies his orchard brings in, he does not hoe it as a *Beschäftigung für Kalikes* (work for cripples, a Yidish expression stressing its unimportance, as a way for the pitiable to pass the time). Michael explains (recorded as he was moving aluminum irrigation pipes in his orchard in the summer of 1977):

It's man's nature everywhere to try to become bourgeois, while socialism was and remains a vision and ideal based on overcoming egoism, that's to say, *"virtu"* (in French), that's to say, giving up some personal benefits to allow society to progress. To maintain the orchard by the work of the members alone [with some hesitation] O.K. . . . perhaps you don't have to do every last thing yourself, but as much as possible, that, for us is a necessity of existence . . . , because, if we don't make the younger generation face, on the one hand, its weaknesses, and on the other, the need for every person to be aware of, and ready to contribute to, improving society, there is little chance of changing things for the better—we can only do that together.

People who don't have the means of facing themselves can't make a society. The trouble is that in the media and in the schools of today [most of the moshav children study in comprehensive vocational schools in the neighboring towns], they teach you to rate people by their clothes, so who am I to tell my children that a "blue collar" is beautiful? Status symbols are taking root in society and setting the tone. Its a broadcast on a different wavelength, in a different code. And today it's not just clothes. Once they used to say *"Kleiden machen Leute"* [clothes make the man—viz, his value], but nowadays, to be exact one ought to say *"Die auto machen Leute"* [the automobile makes the man—viz, his status]. Not to look ridiculous, you need to buy everything and eliminate the "impractical."

The retreat by the youth of Noon Yud from agriculture was influenced, according to this explanation, by its becoming an embarrassing "visiting card" for them. Even though in and of itself the work is enjoyable and good and engaging in it adds to the internal beauty of the community, it spoils their image in the eyes of their friends from outside. Does this mean that it is the status of a tiller of the soil that has been injured and no longer bestows respect on the young people of the maturing generation in the new Israel? (See Ortiz 1979 on the devaluation of agricultural labor.)

We can obtain a partial answer to this question from the point of view expressed by sons of a moshav close to Noon Yud who left a life of plenty with their parents to set up a unique farm under difficult environmental conditions in the Negev desert in southern Israel. They partly expressed their stance during a conversation that I started in January 1988:

MEDIATOR: This is your fourth year here . . . , and if you haven't yet achieved economic stability, at least you've proved that one can make a living on an arid hillside like this. [*A look of astonishment, a shrug of the shoulders from lack of desire to relate to this.*] It makes one respect you because it requires a constancy of purpose and no small measure of self-sacrifice.

ALON: What's that?!

MEDIATOR: There must have been other short-term volunteers, let's say for a month, who came to the area or were ready to work, even by your side, for a certain period, but determination is measured over time, and for four years now you've kept going. It's not been easy for you, but you haven't given up, and now you can begin to see results. You are leaving footprints in the sand. Stamina is also a measure, isn't it?

NURITH: It's not as hard for us here as you think; it's just nice here. [*They notice my embarrassment at having entangled myself by my taking a high tone and come to my aid.*] The moment it stops being a pleasure, we'll just stop and go somewhere else.

MEDIATOR: What things in particular make life here on the hillside so pleasant?

ALON: First of all, life is quiet here. From morning to night, there's nobody to order you about; and another thing, only someone who loves the nature of the region can understand.

MEDIATOR: Alright, but all the same, you're surrounded by desolation, so there are those who would call your putting down a stake in this hill a pioneering act. Don't you feel like pioneers? How is that you're seen as pioneers and you don't care?

ALON: If we had found the quiet and the nature we have here in the center of the country, we wouldn't have come here. Just go near Tel Aviv and see what it's like . . . , [*having second thoughts*].

NURITH: We don't say it's bad in Gush Dan [the name given to the highly populated central coastal region], we just say it's hard to find there a nice spot like this.

MEDIATOR: And what's bad, say, in the hills of Kashmir, the Appalachians, or the Andes? Does the farm have to be situated in Israel?

ALON: We think in Hebrew, and here [in Israel] it's comfortable, although we can get by in English and Spanish. We've no reason not to move somewhere else if it's worth it, why not? We've already toured many countries and laid our eyes on quite a few spots. We've no special obligation to this country . . . , if they were to offer us a place abroad with conditions like here? Sure we'd consider it on its merits. [*after further thought*] But the offer would have to contain no less than we have now.

MEDIATOR: Why are you so against the notion of being called pioneers?

ALON: [*after thinking for a minute and throwing a stone at a dog sniffing around the cattle trough*] First, if you say pioneer, it means that later on others have to follow, right? And we don't want to be flooded. We want to stay here in peace. Second, people place their hopes on a pioneer; he's expected to do certain things, yes? And we don't want to have to do anybody a good turn, or even indirectly glorify or humiliate somebody by means of the life we've chosen. We want to live private lives, obligated only to ourselves, and that's all.

MEDIATOR: So O.K., you live your lives without giving yourselves fancy names. But how can you prevent the interest you are creating? People look at you from outside and see you as accomplishers. There is some vision to which you are connected . . . , does that bother you? Do you want to prevent it?

ALON: Sure! [*with some qualification after further thought*] It's not that we want to or can interfere with the thoughts of others . . . , journalists come and publish articles about us that bring curious people out here in their wake. . . .

NURITH: Afterwards there are those who offer to lend a hand, or are sympathetic and offer us sums of money, which we usually politely decline. . . . The trouble is that we, ourselves, begin to see us as part of a venture; then we begin to complain about government offices we actually don't give two hoots for and want nothing from, and that gives life a bitter taste.

MEDIATOR: Do you follow the anarchist doctrine, that we can do without the state and national institutions? For instance, when, against your wishes, you're bouncing around government offices among the "dumb bureaucrats," what complaints do you make to them?

ALON: To give you an example? The road from Eilat to Tsnifim is excellent. The road from Makhtesh Ramon to Beer Sheva is good. But the section of road from Ma'ale Ha-Meishar to Naḥal Faran and Tsnifim, only a forty kilometer section, is old, full of potholes, and dangerous. A few million are needed for repairs and they'd be able to attract . . . If just five percent of the internal tourism that flows down to Eilat via the Arava road were to return by the hill road and stop off at the Ramat Hanegev tourist sites; the restaurants, the gas station, the guest houses, they'd spend some money and provide employment for hundreds of local families. Such a simple idea to get things moving. . . .

MEDIATOR: It sounds as if you suddenly got turned on by something Zionist, don't you agree?

NURITH: [*self-justifying and wiping an unwanted grin off his face*] It's not that we're against Zionism; we simply want to say that what brought us here was not Zionism . . . Let us tell you, we're nobody's *freiers*!

The young people talking like that are Ashkenazim, in their thirties, and wiry laborers who were educated to lend a hand in the creation of Israel, to make the deserts flourish. They have also learned a chapter in the book of veering winds of ideology; an ideology that sends people out on self-sacrificing missions and then washes its hands of them. The deteriorating status of agriculture is not strange to them, and they are aware of the growing tendency to measure things in terms of material profit. The conflict between Zionism without quotation marks and Zionism with quotation marks they solve paradoxically by performing Zionist deeds, camouflaging their motives with statements suited to the spirit of tough individualism and achieving their vision of personal fulfillment. Their words are similar to those of their own generation in the West, translated into Israeli Hebrew of the 1980s.

## FAR AWAY FROM THE FIELDS

We left Noon Yud, whose sons have extricated themselves from the short-lived tradition that their parents established,[6] understanding that climatic influences holding sway in the surrounding Israeli society have been responsible for the consequences. An analysis of Nurith and Alon's words shows that one cannot talk of a general turning away from farming. One cannot dismiss it as work unsuited to the Jewish nature as some—sickened by the conditions of farming work in moshavim in the center of the country by contrast to the periphery—tried to explain to us. The connection between the external and internal circles are illustrated by the results of a survey of the graduation classes of youngsters from Noon Yud from the years 1978, 1982, and 1986. Tables 10.1, 10.2, and 10.3 list their study programs and their employment or choice of preferred employment for those still studying. Eleven completed their schooling in 1978, seventeen in 1982, and twenty-nine in 1986.

As shown, the tendency is toward the academic and service professions and away from production. What explains the youth's change of mind? Here are some comments, beginning with those of Haim (among the last of the farmers in the village growing vegetables in the open). Haim (forty-five years old) explains the youth's retreat from agriculture as a drastic, rapid development that had already begun in 1968 with the employment in the village of workers from the "territories" (Arab laborers from the West Bank and Gaza who became widely employed in Israel after the Six Day War). Haim opens by making a rebuke:

How are they [our sons] to blame? If I were their age and took a blow like that, I doubt I would choose the occupation I have today. As soon as the workers from the territories came, all the work could be carried out here at low cost by a third party . . . , so why work hard? It occurred so fast that even I didn't grasp what was happening to us . . . and then the Yom Kippur War came and caught us in the middle of the autumn harvest without the workers from Gaza. [During the emergency situation from October 6 to the end of the war, Arab workers generally stayed away from work inside the Green Line.] Nobody had realized how fast we had become accus-

**Table 10.1**
**Trends of Study, Career Tracks, and Present Employment for Noon Yud Graduates, 1978**

| Study program/number | Level of training | Employment |
|---|---|---|
| theoretical - 5 | academic (4) | social worker (1) |
| vocational - 3 | seminar etc. (3) | engineer (1) |
| agricultural - 2 | other (4) | vet(1) |
| incomplete - 1 | | lawyer(1) |
| | | teachers (2) |
| | | paramedic (1) |
| | | housewife (1) |
| | | farmer (1) |
| | | workshop owner (1) |
| | | transport worker (in the U.S.A.) |

**Table 10.2**
**Trends of Study, Career Tracks, and Present Employment for Noon Yud Graduates, 1982**

| Study program/number | Level of training | Employment |
|---|---|---|
| theoretical - 10 | academic (5) | economics (1) |
| vocational - 4 | other (12) | education (1) |
| agricultural - 3 | | biology (1) |
| | | engineering (1) |
| | | computers (1) |
| | | teacher (3) |
| | | farmer (3) |
| | | service worker (2 masseur, waiter) |
| | | mechanic (1) |
| | | insurance agent (1) |
| | | housewife (1) |

**Table 10.3**
**Trends of Study, Career Tracks, and Present Employment for Noon Yud Graduates, 1986**

| Study program/number | Level of training | Employment |
|---|---|---|
| theoretical - 16 | students at academic | studying (12) |
| vocational - 7 | institutions (12) | management and sales (3) |
| agricultural - 5 | others (17) | teaching (3) |
| incomplete - 1 | | full time army service (3) |
| | | nursing (2) |
| | | hotel service (2) |
| | | building worker (1) |
| | | goldsmith (1) |
| | | on a trip abroad (2) |

tomed to that ease and comfort. We had to rely once more on our own work force, so it didn't seem unreasonable to do what had to be done, simply and obviously to "bother" our kids to come and help. . . . In any case, there were no classes and they had nothing to do except to be glued to the radio. . . . So we notified them to assemble at 6:00 AM; we expected between thirty and forty and only nine turned up. Another three girls and a boy arrived at the field at 8:30 AM. . . . Then we talked about money, eight lirot for six hours of work [to midday], and seventeen came. The second day of work was also so-so, and you could still understand it. It takes time for the body to get used to bending over. . . . In six hours they managed what a regular worker does in one; they cracked jokes, ate and drank, went to the side to piss . . . , still, finally you must get used to it, don't you? But with them it got worse from day to day till it became intolerable. When they couldn't handle it any more, you could sense it by the damage they left, pelting each other with produce—"zucchini wars!" They played tag, they ran around and rioted among the rows, uprooted seedlings. . . . In the end we had to kick them the hell out.

Dalya and Ronith, both seventeen years old and pupils of the eleventh grade, were there, and they tell another version of the same story:

The work was fun, until some workers from Gaza had returned, to whom he [Haim] paid by the end of each day. We got used to the physical toil with no problems, and the team, including some grown-ups, was effective and pleasant; but he who was all out thankful to us at the beginning began yelling at us, showing the manners he had acquired talking to his Arab workers, and calling us spoiled brats. And when we talked back to him, to keep us quiet, he began to play on the money string. So far we were all engaged doing voluntary work and no one mentioned money. But if on a payment basis, the little he offered and the trouble he caused us bossing around . . . was no fun any longer.

Picking of the Washington orange begins in November, and back in 1973 the number of Gaza workers at the work sites of Israel was yet small. The local youth were summoned to the harvest; a few arrived, and for a fortnight the team work continued while the man in charge of the grove complained about the meager output, accusing "the kids" of being lazy and careless. Mariana (twenty-seven years old), a housewife and formerly a volunteer from Holland who had joined the team, explains her view of what happened:

As long as the local team was all by itself, it was pleased; the work was done considerably well, with jokes flying around. Being together with cheerful young people, then who cares about mud on the shoes or socks soaked with dew and rain? But then came the Arabs back to work together with a team of men, faces of dummy expression, greedy eyes, hanging jaw, and brown teeth; who wouldn't mind mud on the shoes?

## WHOM CAN WE KEEP UP WITH?

When the unemployed work force from the administered territories began to return to the moshav fields, the departure of the youngsters from the fields and orchards increased. The large number of available workers reduced the cost of simple labor and lowered the status of blue-collar workers, in particular among Jews of oriental extraction and Israeli Arabs (Kressel 1992; Semionov and Lewin-Epstein 1987). Moshavim that were based on arable farming, especially those of the Eastern aliyah (wave of immigration in the early and mid-1950s), and other moshavim that had not established bounds of permitted and forbidden as regarding the principles governing work ethics to help fend off the flood of workers from Gaza, altered beyond recognition. The norm changed insensibly, and false justifications were given for it, as Michael explains: "It's preferable and right, taking the social dimension of the problem into account [since we are presently responsible for the welfare in Gaza], not to take away from them [the Arab workers] the simple jobs both they and we can do, and for us to do the work that only we can."

In the beginning, operating tractors was included in the work members carried out, but later on that too was given over to skilled Arab hands. In parallel, it became necessary not to be seen toiling over "dumb" physical work more than was essential; as one of the veteran members, a graduate of a youth movement in Argentina and a former kibbutznik, contemptuously said of the moshav youngsters: "As for them, they mustn't make a mistake and exert themselves [working] during the day whenever it's possible to save a little energy for the evening."

As they describe it, overseeing the work done by the Arab laborers is necessary, but one must take care to do so sitting in the driver's seat of your car. In the fifth year of the existence of the Green Line, not only the operation of mechanical equipment but also the overseeing of the Arab laborers from the driver's seat has been largely handed over to Arab subcontractors; and since they have been seen driving around the moshav on tractors, accepted conventions have changed. Thus, the members' own habit of using tractors around the village to run their daily errands has faded—it's too much "like the Arabs." From this point on, use of the Ferguson has been left to the subcontractor, while the members, even for the shortest of local journeys, use the car. The question then arises: What is the force caused by the presence of the Arabs that so repels their Jewish colleagues? One Saturday evening, a group of members discussed the subject (recorded in the winter of 1980); Akhinoam, a thirty-eight-year-old resident of Noon Yud and a nurse in the local hospital, expresses her opinion and explains an important point:

AKHINOAM: In the hospital, side by side with us, there are dozens of Arab doctors and nurses who come from the local towns and also from villages in the north, and so far there has been no special problem accepting them as colleagues. It's a plain fact that the Jewish teams don't switch their work because of them.

MEDIATOR: Perhaps what counts in the hospital is their relatively small number [compared to the number of Jewish workers]; as long as there are few of them, it doesn't matter, because their participation doesn't worry the others.

AKHINOAM: It's true there are hardly any Arab nurses, but with the sanitary workers, who are mostly Arab women, we also have a large total number of Arabs.

MEDIATOR: Then perhaps it's because the respect accorded to doctors and nurses is guaranteed; perhaps the good salary makes the difference. If you don't mind, let's take an opposite case; suppose the demand for unskilled workers was so great that they offered, say, to fruit pickers, the wages they give a doctor. Would that be enough to increase the willingness of the village youngsters to work in their company, to see them as colleagues?

AKHINOAM: If, for instance, the heads of the hospital departments were clever and talented Arabs? A good question. If they were the modern, cultured people and the Jews were the trainees, then perhaps they'd be like the Jews and we'd be like the Arabs. [*general laughter*] Then we'd surely be the ones seeking their company, and maybe they would avoid us.

MEDIATOR: If that's so, perhaps it's easier to accept the Israeli Arabs as colleagues, especially those who have made progress, while the Arabs from the territories who are relatively more primitive are harder to accept?

The question is answered by Yavin, a forty-five-year-old village member and manager of the large regional citrus packing plant:

YAVIN: You'll be surprised to hear that quite a few of the fruit pickers and building workers are high school graduates and even have technical college diplomas from Egypt and goodness knows where else.

MEDIATOR: Then it's likely that the problem of their presence is not actually in the bad name it brings to the job, as it were, of work for the unfortunate but in the climate, work relations, and the sometimes unpleasant friction.

YAVIN: If there's a retreat from farming, that's not yet a sign of friction. It may be the opposite, that work relations are based on a clearly defined ladder of authority that doesn't allow any clashes. What may be and is probably at the heart of the matter is the rotten feeling those in authority have working with them, chiefly the Gazans. Just thinking about this Gaza, a bottomless pit of problems, makes you want to run a mile.

This last explanation for the retreat from farming in the village highlights what the presence of the Gazans (in particular) on the farms has done to the connection between the member and himself. That is, it is not simply a question of the depreciation they have brought to the job but the touch of guilt that penetrates the members' hearts when they put them to work and see themselves as wrongdoers (Semionov and Lewin-Epstein 1987). This trait perhaps explains how the subcontracting of farming on the moshav passed so quickly into the hands of Arabs.

Since the company of the workers from the territories is not that unpleasant for all the members (there is the motivation to work with them), we tested another explanation offered by the forum—the infiltration of their (the Arabs) traditional system of values, which conceals a contempt for manual labor, into the local circle of members that worked with them. Haim explains:

For me, working was in my blood. During the first period [of employing Gazans] I worked with them at everything, but right away you see it doesn't improve your standing in their eyes. It's not like with us [the industrious are better than the lazy]. They uneasily hint that it's not necessary . . . , when I used to load the crates, my *rais* [foreman] would take them from my arms saying: *"mush lazem! mush lazem!"* [not necessary that you tire yourself out] . . . It's not that they just want to curry favor, they simply want to see you as the boss, and a boss who sweats harms his status, and when you go down in their eyes, you endanger your authority.

The value system of the Arabs from the territories is a matter for specific research. If, however, their representatives working at the various Noon Yud sites cast their cultural shadow over the Hebrew work ethic, it is worthwhile testing the question, How did it happen so very easily?

Here is a description of talk that we recorded at one of the moshavim in the Negev that forcefully raises some points not talked of in the above discussion at Noon Yud. Eli, a fifty-year-old Tunisian from a moshav with members from a variety of family origins and owner of a fruit-tree farm and a herd of sheep, explains how the entrance of Gazan workers began and what their influence was (recorded in 1988):

Workers from the [Gaza] Strip first came for the citrus picking in the co-operative orchard, directed to us by the labor exchanges, in the winter of 1969. Our contact man with them was Moshe Mizra'i from the second generation [in Israel] of immigrants from Syria who was fluent in Arabic. After the citrus harvest, he kept a few of them on to pick apricots in his own plot, and actually, from then on they stayed with him and not just for seasonal work. . . . At the start, there was some criticism, mainly because at the end of the day he'd let them sleep in sheds in the courtyard, but all in all, it wasn't a bad arrangement. Moshe was able to supply workers to whoever had urgent need of another pair of hands and became a big man. When the matter was raised at the general meeting [winter of 1971], there were already a number of farms with sub-tenants like that. At first they still looked pretty shabby and weren't very noticeable. There were those at the meeting who called for their expulsion, while a few took pity on them.

The Arabs slowly learned more and more from Moshe, and those who picked up some Hebrew no longer needed a translator. The presence of this reserve of "assistants" quickly spread, and those of the members who resisted employing them, began to feel they were being left behind. In some bitterness, they asked for the intervention of the Movement, transferred the matter to what we call the "Institutions," and under their wing tried to take up the fight again at home. But it was already too late to change the situation. They remained a minority, not just in our moshav. Whoever carried on doing the work themselves became exhausted. Perhaps the moshavim of the Arava [the rift valley running from the Dead Sea down to Eilat], like Ein Yahav, which don't have the "administered territories" next to them, can play at "principles," but here they [Arabs looking for work] are swarming everywhere.

## DEGRADING EVERY FORM OF MANUAL LABOR

In 1989, two of the village youth, waiting to be called up for their army service, took on the job of cutting lawns. Another two together raised a field of fodder that they sold after its harvest to horse breeders in the neighboring village, making a nice profit. Another four worked as shop assistants in local towns, three were counselors in children's summer camps, while the rest (the majority) waited for the call-up and did nothing. These few representatives of the large graduation class, who could be seen every day in their work clothes, focused on their work and, pleased with themselves, were approvingly received by their parents and their friends' parents. They were produced as evidence to support the statement, "Look, it's possible, the chain has not been broken."

Others were less impressed and preferred to mourn the spirit of the past. Pessimism over the chances of the village surviving as a village, which I had already recorded in 1978, reflects this mood. Dov, an eighty-year-old father of five and grandfather of twelve from Bielorussia who was raised in Argentina and immigrated in 1951 concludes:

They're not the reason [the Gazans]; it's our politicians, from the Jewish streets. They [the politicians] go up and down between Jerusalem and Gush Dan all the time, passing by us just a couple of meters away, but they never stop. Maximum, they stop off at the Ramle market where the [agricultural] produce is good and cheap. . . . What would happen to them if they were to spend a little time in an orchard picking fruit? So people should see—they know well enough how to bring along a photographer and a journalist to preserve them for posterity, putting on a pair of boots for a little fruit picking! That would influence the youngsters' scale of values more than a hundred Arabs. Everybody should look, the Arabs too, at the Prime Minister or one of the members of the cabinet putting in a little time in work clothes. What would be wrong with that? Then a negative label for such beautiful work wouldn't be possible. All we've got is a Knesset (parliament) of white collars and ties, lawyers from Tel Aviv with lily-white hands like fine gentlemen. . . . Look, I'll tell you about my kids, something that happened a month back; they were quarreling about who was going to roll up his sleeves and, actually, who wouldn't lend a hand to finish putting in the drip irrigation in the orchard. I had bought the materials [plastic tubing and drippers] and left it beside the trees, ready for them to come. Earlier, we'd talked about a "special effort"; all three of them would come, with their children, for a weekend of hard work, once again together with their Dad, like when they were growing up. Well, they arrived and they hadn't even gotten out of the car before they were fighting about who should do more and who should do less.[7]

Finally they got out of the cars, drank some coffee, ate some cake, and the time came to roll up their sleeves. So they did that and got a little done, a very little. Most of the "special effort" they left to the granddad, to somehow get it done with the Arab workers. After all, they were educated to the work, were always energetic, ready for any effort. At first, after they went away and settled into work in the city, they used to come back at weekends to help on the farm, they used to say they missed working outside in the open air . . . , and now, psshh . . . an attitude like that, where does it come from? Perhaps they're embarrassed in front of their wives and children, or their new city friends. In the city, a game of tennis is the acceptable way of "stretching your bones," and for village air you go on a picnic; and a Jew who still works the fields is a "sad sack," someone who hasn't found a way to get out of it [farming] in time, someone who, at best, is a "nature preserve" [staying on from idealism].

The feeling that the authorities are responsible for national morale and the spirit of the next generation is shared by most of the members. The criticism of the unchecked flow (or as they called it, "nonstop policy") of workers from the territories to the moshavim was also linked to the activities (or failures) of the government. On the basis of the understanding I gained from Noon Yud, I formulated this question: In what way, in your opinion, is the government failing, in view of the resulting attitude of the moshavim to work? This question was put to moshav members in the western Negev by student researchers with whom I worked from 1986 to 1989.

The answer given by David (a forty-six-year-old born in Romania, graduate of Youth Aliyah and member of a Negev moshav from the 1960s) was recorded in the winter of 1988, and in essence it agrees with Dov. He also believes that the behavior of the younger generation, avoiding responsibility for the village fields, is influenced by a public climate of evading responsibility for public (national) property evident, he says, throughout the country. This is influenced, he believes, by a lack of personal example from the country's leaders. However, his point of view is different in his identification of the guilty leaders and in his demands that they correct the situation:

DAVID: It's all a matter of education, and just as you can compete over how much output you can accomplish at work, you can also compete over your achievements in not working. It all depends on the agreed rules of the game, and I had the opportunity to learn that immediately on my arrival in this country. I was fourteen years old. I had no family and I was sent to the Ahava' orphanage near Haifa. . . . There, the education emphasized love of work according to Gordon's gospel [a famous pre-State socialist man of distinction] to the extent, for example, that the undisciplined kids were punished by taking away their right to work.

Once, they caught one of our group pinching some stuff from the "economia" [storage cupboard]. His punishment was particularly charming in its phrasing, "removal of the privilege of working [on the institution's farm] for a whole week." We received it with sad laughter, as something serious and at the same time entertaining. It took us some time to learn how to react to the attitudes of the teachers with an independent mind of our own. . . . In a spirit of *dafka*[8] as a game, we set up a "rule book for energy savers." We imposed on ourselves bans and boycotts for the crime of unjustified wasting of energy. And we instituted half-mocking punishments for anyone caught breaking the norm and making too much of an effort. Like, for instance, going to the dining hall was a must, you had no option, but to make the effort to turn the door knob or push the door open, we decided, was excessive. So we'd stand near the door to the hall, pretending to read the notice board, till somebody would go in and then we'd quickly squeeze in behind them. They had a kind of springy swing doors there that would swing in and out till they settled down.

Another method was to accompany some girl [the group excluded girls from the secret roll call], who would innocently push the door open for herself and us. A kind of sport—the main thing was not to touch the knob, to enter without taking one's hands out of one's pockets. Once we caught a boy called Hilik not restraining himself, opening the door and running to eat. His penalty was to schlep Isaac on his back, like a sack of flour, all the way back to the dormitory. You might say we expended vast energy for the saving of negligible energy. But those were the rules of our game. We were lucky, of course, that we were not required to slave in order to live. They gave us all our meals in the dining hall without considering if, how, or how much, we worked . . . , when you're supplied from the horn of plenty, it's easier to develop games of taboo.

MEDIATOR: Are you saying that the comfort in which the children are living spoils them to the extent that they escape to a land of games, a land of dreams?

DAVID: From newspapers and books you can learn that even on the poverty line, *clochard* cultures develop, of conscious, born beggars, who sleep in the refuse, but will do everything not to be ashamed and to seem to be working. It's a matter of outlook, but . . .

MEDIATOR: You are really saying that perception is what actually determines experience and not the other way around, as Karl Marx put it?

DAVID: Everyone's perception here can determine his own experience, even to the point of death. And what brings people to suicide if not a state of perception? No, but what I'm saying is that education will determine how we live here and how . . . , the education I got at the orphanage won in the end, and I bring up the matter in connection with our kids. It was the orphanage that taught me to handle tools, got me used to the personal discipline of getting up and going to work, of a full day's work, and I know today that not everyone who's inured to manual labor has a hankering for physical work. . . . But there are those who get genuine satisfaction from doing something well with their hands, like a moshavnik who is still skillful in harvesting with a scythe, or a carpenter whose profession is right there in his hands. A carpenter like that will feel he's withering away if he can't feel his tools in his hands. He'll thank you for letting him go to his workbench—but not at the price of the contempt of his environment. The woman is the barometer. If she forbids her husband to carry on, he'll give in for the sake of peace at home. Healthy youngsters are ready to climb mountains, to tour the world with a backpack . . . , my eldest has a horse that he looks after with much effort and goes riding with his friends in his spare time, while my second goes yachting till he's exhausted. In all, they're not lazy, but if they're going to work up a sweat, they won't do it to for "work."

MEDIATOR: If that's so, what do you want of the politicians? Were they the ones who changed the rules of the game, like your "energy savers rule book"? I'm sure some of them toil away hard enough at their job.

DAVID: They've got to understand that studying for agriculture, or for vocational training must be part of the general education of every youngster in Israel. Just as it's important to impart knowledge of Judaism, and how to handle a weapon, it's important to impart the perception that we cannot survive here except by our own labors; that every high school graduate and every doctor can tend his own garden with his own hands, fix up furniture in his house, repair an electrical switch, solve a plumbing problem. . . . After we took the leap from a Diaspora mentality, a people of *Luftgeschäften*, to a mentality of nation-building, we've got to secure what we have achieved. . . . When, in the educational system, vocational training is left to those who fail academic studies, when they brainwash us with American movies full of "beautiful people" with time to spare, and hardly anything about life in the country or what goes on in factories . . . that way one can forget everything we renewed in this country, the mentality of our old lifestyles . . . that's the way we're letting the Diaspora saturate our country again.

Here the criticism of the leadership focuses on pedagogy. Its essence is the lack of care in education for values essential for the national morale. Reinforcement of this criticism comes from Shaul, a forty-nine-year-old Tripolitanian from a religious moshav in the western Negev who blames the authorities for not controlling television programming:

It [turning your back on farming] comes from outside, not from inside. Western ideas have taken over the country. In Western Europe and the U.S.A. nobody cares for work anymore. They have the Blacks do the work for them. They bring in Blacks for any unpleasant job. Just pay and sit back in comfort. Bow down to money! Our kids imbibe that from the media. It drips into the mind from what Israeli television broadcasts. It's all entertainment . . . , that's what decides for our youngsters what they want to be when they grow up. Look, for instance, my grandson, aged four and a half, loves helping me feed the chickens in the afternoon. I used to take him with me until my son-in-law [twenty-seven years old], also from the moshav, came up to me and demanded that I leave the boy alone to watch children's television programs. I answered in my own way, all honor to the television, but that it was also important to help him love looking after animals. To that, he answered, "What do you want, that we should raise the children the way you raised us?" and he made what he meant quite clear; to work all day like an ass and be a cheapskate all your life!

This version concludes that the "laws of osmosis," which also apply to life in society, do not permit the existence of small isolated cells such as a moshav community. The cell walls are permeated by the surrounding Israeli environmental influences, and parents want their children to move up to more prestigious and better paying professions than their own.

During 1991, as a result of the *intifada* and the scarcity of Arab farm workers and, at one and the same time, the major influx of immigrants and the unwillingness of both immigrants and locals to take on the status of wage earners in agriculture, the Ministries of Labor and Finance announced their readiness "to compensate workers directed into agricultural labor, with a sum of 15.00 sheqels per day over and above the wages paid by the farmer." The effect of this step on the placement of workers in Noon Yud's fields and orchards remains limited. During the past year, members of the cooperative have won their battle to rezone agricultural land for the building of houses for the younger generation and thus have signaled their intentions of slowly divorcing themselves from farming.

## CONCLUSION

The sending of immigrants to the moshav and the encouragement of a movement from town to country, were both short-term arrangements in opposition to the modern worldwide trend of drastic reduction in the number of farmers. Development of means of production and investment in modern technology continued apace even though a plentiful supply of agricultural produce to the local markets had been achieved (at the end of the 1950s) while attempts to widen Israeli export markets were meeting with difficulties. From then on, the numbers who could live comfortably from farming dwindled, and competition between them increased. The more highly mechanized and efficient survived.

To meet the need to settle the country and to distribute the population widely, it was the government's policy to direct farm subsidies to the moshavim on the periphery and to deny them to those in the central area. The result was the exit of many farmers from the circle of production.

The Moshav Law, which prevents the splitting up of the family farm, and, according to which, only one of the children can inherit, acted to direct the other children from infancy either to establish their own farms or to find alternative employment outside agriculture. Leaving the moshav to go to work increased with the development of various industries and services in the country and with the proliferation of the bureaucracies. External employees who continued to live in the village and other settlers from outside contributed to its change in character and transformation into a suburb. Most of the farming plots that remained were rapidly handed over for cultivation to Arab workers or a few Arab subcontractors.

A desire for increased efficiency led the farmer to move from labor intensive crops (such as vegetables) to orchards, compatible with outsider labor, but only at harvest time.

From 1968, the cost of manual labor was significantly reduced by the availability of Arab workers from the territories, and this lowered the status of blue-collar workers on the moshav. At the same time, the desire to avoid the negative label attached to employers of Arabs became a feature that characterized, in particular, moshavim of the western (Ashkenazi) aliyah.

The alternative answers found to the challenge in the various production areas were in essence these: (1) to avoid crops needing much back-bending labor[9] in favor of other high-tech crops; (2) to avoid agriculture and work on the moshav altogether; (3) and to ignore critics and avoid self-criticism, to "bend with the wind," and thus to profit from the labor of workers from the territories. The choice was not easy for the following reasons: (1) To develop alternative, high-tech crops requires considerable capital investment, while the volatile cost of credit (typified by the situation at the end of 1985) can intensify the difficulties of the borrowers. (2) Branches that are more or less impermeable to the entrance of Arab workers are the civil and national services, in particular the defense forces and the police, and also defense and high-tech industries. However, one requires professional training and talent to be employed by them. (3) Employment of Arabs impairs the motivation to work and erodes the justification of the moshav ethic and personal pride.

It has been years since agriculture ceased to be the source of income for most of the families in the village. It is apparent that the philosophical links that were the initial basis of attitudes toward it have also vanished. This is heartbreaking for the veterans, all the more so since the younger generation's attitude of turning one's back on any form of physical labor has deepened. However, the placement of the children in alternative employment has been as much influenced by the families' desire to guarantee their future than by the surrounding Israeli cultural climate that has begun to reject the dream of national renewal in the homeland.

The use of Gazan workers in Noon Yud farms hampered the attractiveness of agriculture for the young and ambitious moshav members. The reasons for this are the following: First, since the time most of the rough work was deposited in the hands of Gazans, farming has been losing its "sportive value." Demonstration of individual ability, as in competitive games, must follow clear rules which should obligate all participants equally. The quantity and quality of production by the end of a day of work or an agricultural season resembles the results of a sprint at the finishing line. The entrance of Gazans into the farms of the less sportive farmers could not leave their decently competing colleagues indifferent (see Eli's words). Eventually, all of those who went on farming had to thus employ Gazans or quit. As in long-distance running, we find an analogous image of a retired farmer in the eyes of his maturing son (Dani's confession) who consequently shuns farming. Second, socialist consciousness (paying a great deal for maintenance of cooperative institutions, thus paying homage to egalitarianism), which had given a collective significance to individual work, compensating for hardships and lack of comfort in moshav's life was of no avail once Gazans were let in to fulfill the dirty jobs (see Michael's words and Yavin's words). Third, nor did striking the Zionist chord (the words of Nurith and Alon) lead to an understanding that people were getting their living the hard way for the sake of a collective goal (i.e., nation-building).

Attempts to instill competition among teenagers over the performance of work quotas (see Dalya and Sigalith's reports) began when "big [Zionist] words" lost their effect. The offer of extra payments as a means to spur competition or just increase the output in orange picking by taunting the feebleness of workers as their shoes get soaked in mud had an adverse reaction (Mariana's report) because as long as the Gazans were present, money could not make a difference. Why did moshav members rule themselves out of competition with the Gazans?

*Homo competitivus* searches for tracks where ability can be shown, be it through achievement or through resolute laziness and the saving of body energy (David's story). Speeding at work is motivated by the income the greater output can render, as well as by the fashion according to which the income can be spent. Spending and saving serve also as tracks for competing, and interestingly, these two alternatives are interrelated. We find that the choice of vocation is made not only in consideration of its income, but also in the conventions of its work effort, creativity, quotas of production, and in the standard of living such a salary may afford. The image of one's future colleagues and their standard of living or the fashion by which they spend the salary are also important. To this way of thinking, cheating as a means of winning a game robs the action of its play-character and spoils it altogether because for the actors, the essence of play is that the rules be kept—that it be fair play (Huizinga 1950). This aspect should be added to the various factors appearing in the literature pertaining to choice of vocation (Lissak 1969; Holland 1973; Ortiz 1979; Wallman 1979a, 1979b; Super 1980; Krue 1983; Nevo 1987).

In accordance with this analysis, the Gazans and their preferences for spending the work revenues earned in Israel affect the way their occupations are looked upon. The apparel of the Gazans at the workplaces and at their homes now color the jobs in a way that moshav members do not want to be painted.

## NOTES

1. The term appeared with the effort to make the Jews return to primary occupations, such as the labor on the land, with early Zionism.
2. On the interplay of winning through cooperation, see Orlick 1978.
3. See Schwartz and Giladi 1993.
4. Dov Baer Borochov (1891–1919), the first theorist of Socialist Zionism, advocated the "normalization" of the Jewish vocational structure by the turning of manpower from secondary occupations (craftmanship, trade) to primary ones (farming, mining).
5. As for the use and the implications of the term *freier* in Israel, see Roniger and Feige 1992.
6. For parents having children and then helping their absorption outside the moshav in the Western Negev, see Shokeid 1990.
7. I asked about the details of the argument, and here is the explanation: The exact topic was who would excavate the channel with the digger to lay down a section of two-inch pipe across the exit road between the plots. There the ground was harder and relatively difficult to break up, and the second son (thirty-nine years old) said it was nothing—a matter of half an hour or so. However, on principle he was waiting for his older brother (forty-one years old) to go first for once and dig the hard section because the last time they had both come for fruit picking, the older brother had taken it upon himself to stay behind and load the crates for dispatch on the tender (having come late) but just then remembered he had to hurry back to town; so his brother (with his brother-in-law, thirty-seven years old) had to hang on and do that job too.
8. An untranslatable word with many shades of meaning depending on the context. Here it means something like "just because!"
9. Now, for the first time, we record expressions such as, "angle crops" (literal translation) for the culturing and harvesting in vegetable gardens rising to a level of bitterness in nicknames like "Boazim" (Boazim, the owner of the fields in which Ruth the Moabite gleaned) given by the younger generation to the moshav farmers employing Arab workers.

## REFERENCES

Applebaum, L. 1974. *Not by Bread Alone: Continuation and Recompense in Nahalal.* Rehovot: Center for Rural and Urban Settlement Research (Hebrew).

———. 1979. *Moshav HaOvdim in the Test of Time: Forms of Organizational Change.* Rehovot: Center for Rural and Urban Settlement Research (Hebrew).

———. 1983. *The Non-Agricultural Village in Israel as a Developmental Tool: Background, Problems and Chances.* Rehovot: Center for Rural and Urban Settlement Research (Hebrew).

———. 1986. "Migration of Urban Families to Moshavim." Ph.D. diss. Jerusalem: The Hebrew University.

Grossman, D., ed. 1987. *Between Yarkon and Ayalon.* Ramat Gan: University of Bar-Ilan Press (Hebrew).

Gvati, H. 1981. *Hundred Years of Settlement: The History of the Jewish Settlement in Eretz Yisrael.* Tel Aviv: Hakibbutz Hameuhad (Hebrew).

Holland, J. L. 1973. *Making Vocational Choices: A Theory of Careers.* Englewood Cliffs, N.J.: Prentice-Hall.

Huizinga, J. 1950. *Homo Ludens: A Study of the Play-Elements in Culture.* Boston: Beacon Press.

Kressel, G. M. 1992. *Descent through Males.* Wiesbaden: Otto Harrassowitz.

Krue, E. 1983. "The Vocational Side of a New Start in Life: A Career Model of Immigrants." *Journal of Vocational Behavior* 23: 270–285.

Lissak, M. 1969. *Social Mobility in Israel Society.* Jerusalem: Israel University Press.

Nevo, O. 1987. "Irrational Expectations in Career Counseling and Their Confronting." *Career Development Quarterly* 35: 239–250.

Orlick, T. 1978. *Winning through Cooperation.* Washington, D.C.: Hawkins and Associates.

Ortiz, S. 1979. "Labour and Value among Paez Farmers." In *Social Anthropology of Work*, S. Wallman, ed., pp. 207–208. London: Academic Press.

Roniger, L., and M. Feige. 1992. "From Pioneer to Freier: The Changing Models of Generalized Exchange in Israel." *Archives Européennes de Sociologie* 33(1): 280–307.

Schwartz, M., and D. Giladi. 1993. "25% Agriculturalists? Reality and Conceptions during the Formative Decade of Israeli Agriculture: 1949–1959." *Economic Quarterly* 40(3): 391–414.

Semionov, M., and N. Lewin-Epstein. 1987. *Hewers of Wood and Drawers of Water: Noncitizen Arabs in the Israeli Labor Market.* Ithaca: ILR Press, Cornell University.

Shokeid, M. 1990. "Generations Divorced: The Mutation of Familism among Atlas Mountains Immigrants in Israel." *Anthropological Quarterly* 63(2): 76–89.

Shoresh, D. 1988. *Farming Differentiation in the Workers Moshavim.* Rehovot: Center for Rural and Urban Settlement Research (Hebrew).

Super, D. E. 1980. "Life-Span Approach to Career Development." *Journal of Vocational Behavior* 16: 282–298.

Wallman, S., ed. 1979a. *Ethnicity at Work*. London: Macmillan.

———. 1979b. *Social Anthropology of Work*. London: Academic Press.

Zussman, P. 1988. *Individual Behavior and Social Choice in a Cooperative Settlement: The Theory and Practice of the Israeli Moshav*. Jerusalem: Magnes Press, The Hebrew University.

# Chapter 11

# Moshav Ranen Is Going Back to Agriculture

*David Sivan*

*Moshav Ranen seems to contradict the trend observed in the majority of moshavim. According to David Sivan, its members have found agriculture more attractive in recent years rather than less. Not surprising, they have strengthened the cooperative institutions of the community. Sivan attributes the distinctiveness of this moshav to its ethnic and religious characteristics. Ranen was settled by Karaite Jews from Egypt. Karaite beliefs emphasize conduct conducive to economic cooperation under the conditions the moshav has been facing in current times, and membership in the sect has promoted close social ties and mutual assistance.*

The crisis of the mid-1980s reduced agricultural employment (Haruvi, Fleischer, and Eger 1989; Kislev, Lerman, and Zussman 1989; Applebaum, Shoresh, Eger, and Hartman 1990) and weakened cooperative associations. In Moshav Ranen, the opposite is occurring: People are going back to agriculture, and the prestige and power of the Association (aguda) have increased. The purpose of this study is to find out why Ranen differs from the general pattern in these respects and how the two processes are related. The affiliation of all members of the moshav with the Karaite sect is apparently an important factor in its development. Therefore, I briefly review the history the Karaites and that of Moshav Ranen.

## THE KARAITES: BACKGROUND AND CHARACTERISTICS

The Karaites are a Jewish sect—opposed to mainstream Judaism, the rabbinical tradition—and consider the scriptures to be the sole legitimate source of religious law. They reject the Oral Law, for example, the *Mishnah*, the *Talmud*, and the *Shulhan Aruch* (*Encyclopaedia Judaica* 1972). The Karaites also refuse to recognize the permanent authority of any religious leader and emphasize the freedom of each individual to interpret the Bible (religious individualism). The following principles guide the determination of the law (Dvir 1990; *Encyclopaedia Judaica* 1972): (1) the literal meaning of the biblical text (*ketav*); (2) the consensus of the community (*edah*); (3) the conclusions derived from the scriptures by logical deduction (*hekkesh*); and (4) knowledge based on human reason and intelligence (*hkhmet ladaat*).

According to Algamil (1981), the Jewish Karaite sect began to take shape within the Sadducees under the leadership of Yehuda Ben-Tebai (the president of the Sanhedrin) in the first century B.C. during the period of the Second Temple. Others say that the Karaite split began in Babylonia under the leadership of Anan Ben David in the eighth century A.D. during the time of the Gaonim (Hebrew *Ga'on*, pl. *Ge'onim*; "eminence" in English, referring to eminent Jewish scholars of that time) (Colligan 1980; Nemoy 1952).

Karaism developed in response to the depressed situation of the Israelites in the Arab (Muslim) Caliphate in the Middle East (Mahler 1949; Ankori 1959; *Encyclopaedia Judaica* 1972). "The passage of time and the harsh necessities of reality eventually forced upon Karaism a modification of "its social-political principles" (Nemoy 1952). Thus, there remained only the religious peculiarities that underwent modification.

In the beginning, the Karaites lived in peace with the Rabbinic Jews. However, in the tenth century, a decisive stimulus for the consolidation of the sect was provided by the famous Rabbinic scholar Sa'adiah Ben Joseph al-Fayyjmi Gaon, who viewed the peaceful relationship between the Karaites and the Rabbanites as a danger to Judaism. He proceeded "to attack and refute the doctrine and practice of Karaism and to demonstrate . . . that the Karaites were not just harmless deviates . . . but complete apostates who should be ostracized from the community of Israel" (Nemoy 1952). The attack drove the Karaites to increase their scholarly and intellectual creativity, which further consolidated the sect (Nemoy 1952; Ankori 1959).

The sect was spread throughout the Arab empire, while its spiritual-creative center moved to Jerusalem (Nemoy 1952). By the end of the tenth century, there were Karaites (and other Jews) in the Balkans under Byzantine rule. Toward the end of the eleventh century, the center of activities moved into the Ottoman Empire and into Egypt (Nemoy 1952; *Encyclopaedia Judaica* 1972). Records show that by the twelfth century, Karaites were living in the Crimean Peninsula, and by the fourteenth century in Lithuania and other parts of Eastern Europe. The progressive decline of the Ottoman Empire from the sixteenth to eighteenth centuries resulted in a process by which the Russian Karaites assumed the leadership of the sect (*Encyclopaedia Judaica* 1972; Nemoy 1952). Consolidation of the Russian Empire, division of Poland, and the emigration of Lithuanian Karaites to the Crimean Peninsula in the second half of the eighteenth century enforced this process. The emergence of Zionism brought the Egyptian Karaites to the fore.

The Karaites are a religious minority. Weber (1958) pointed out that such minorities try to gain recognition through success in business and thus concentrate in entrepreneurial economic activities. Also, their cohesion tends to strengthen during difficult times.

Ankori (1959) points out that "Karaism stands out as a unique phenomenon," as the only dissenting group which has survived within Judaism. A reason for this might be the flexibility of Karaites and their ability to improvise solutions to the changing reality (Ankori 1959). Indeed, the wealthy Karaites sometimes exerted influence on rulers as, for example, they did in Egypt of the eleventh and twelfth centuries. Some were landowners, others merchants, and still others were involved in white-collar occupations (Corinaldi 1984).

The Karaites' economic prosperity continued into the twentieth century. In Egypt, many of them were engaged in white-collar occupations; others were gold/silversmiths and gold/silver merchants (Algamil 1985). Most of Egypt's Karaites resided in their own neighborhoods in Cairo. They worked and lived together and constituted a cohesive community. The increase in the activities of the Muslim Brothers in 1948 contributed to the feeling of insecurity among the Jews, to the strengthening of Zionist activities, and as a result to emigration to Israel (Algamil 1985). The Karaite group that settled in Ranen came out of this environment.

## MOSHAV RANEN

Moshav Ranen is situated in the northern Negev region. It was established in 1950, as were many of the neighboring moshavim, by eighty families who came from Yemen. Within several months, these families left, and a group of twenty-eight Karaite families recently immigrated from Egypt replaced them. However, the edge of the desert was a harsh environment, and jobs were scarce. Thus, many who passed through Ranen, the Karaites included, did not stay, and the number of families grew slowly.

Following the Sinai War of 1956 and the expulsion of Jews from Egypt, Ranen absorbed a wave of families, and the overall population stabilized around forty families. After the Six Day War, the number rose to about sixty families. By the end of the 1960s, Karaite families were the only ones left. At the same time, the second generation started to join its parents on the moshav. By 1978, all the eighty originally planned units were filled, and since then no units have stayed unoccupied for long.

Coming from the small cohesive community of Cairo Karaites, the settlers of Ranen knew and understood each other. This helped consolidate the moshav community. During the Yom Kippur War (1973) when one of the members was drafted into the army for a long period, the others helped run his farm. And when a member needed an expensive medical operation, the others paid for it.

In the beginning, agricultural activities were carried out collectively. As the settlers acquired experience, they developed independent initiatives in cultivating field crops (i.e., wheat, barley, sunflower, and so on). Within several years, the Settlement Department of the Jewish Agency began investing in the development of land for the different field crops. Each farm unit received three irrigated dunams for growing vegetables. In the mid-1960s, on the initiative of the Settlement Department, seven sheep ranches and twenty-three chicken coops for egg production were set up. By 1973, the sheep ranches proved uneconomical and were liquidated with the assistance of the Executive Committee of the moshav. Every sheep rancher sold his animals and deposited the proceeds with the Association (for future development). The debts from other assets of the branch (equipment, buildings) were estimated, and the sheep ranchers received an interest-free loan (equal to the value of those other assets) for five years from the Association. It should be noted that it was the Executive Committee of the moshav that initiated the liquidation process and, in effect, organized and managed it.

At the same time, following a crisis in the chicken branch, the number of egg production units decreased. They were replaced by fryer chicken units. In the 1970s, hundreds of dunams of orchards were planted on the members' initiative, while the growing of vegetables and field crops continued. In these years, there was also a short-lived attempt to grow carnations in greenhouses. This time the failure was due to competition from African growers that made prices cheaper on the European market. The Executive Committee of Ranen met the crisis in the same manner as it had done with the sheep ranches. In the first half of the 1980s, during the period of high inflation, economic development came to a halt, except for the continued expansion of the orchards to more than 1,000 dunams. At the same time, field crop cultivation began to decline. The institutions of the moshav also played a major role toward the Consolidation (the discontinuation of the patronage of the Settlement Department) and the injection of the Consolidation budget. Parallel with the Consolidation the moshav institutions increased their involvement in the infrastructure and economy of the moshav; Pinchas Gal (from a neighboring moshav) contributed to this development in his capacity as treasurer-secretary and Mordechai Ovadia (a Ranen native and a member) as the business manager. They guided the moshav's institutions in the development process. Parallel to phasing out the responsibilities of the settlement organizations, the settlers of Ranen were wise enough to allocate means to cope with the new reality.

The most significant changes in the branch structure of Ranen came in the wake of branch crises. The sheep and egg branches brought about the initiatives to develop the fryer chicken and carnation branches. When carnation greenhouses were liquidated in the middle of the 1970s, a significant expansion of the orchards had already begun. The branch assortment of today is the accumulated residue of these developments through the years. Moshav members who succeeded in their chosen branch expanded their activities, while those who experienced a crisis moved on to a new agro-economic activity with the assistance of the Executive Committee. According to one of its long-time members, mutual trust, mutual assistance, mutual responsibility, and social cohesion (which strengthened in time of crisis) were the factors that enabled Ranen to go through these changes with few difficulties.

The crisis in the agricultural sector in 1985 accelerated the disintegration of cooperative institutions in many of the moshavim of the Negev. Four factors contributed to this result: (1) Due to the basic mutual trust toward the cooperative institutions, members were willing to sign loan guarantees whenever required. Moreover, Pinchas Gal was able to secure credit (at a reasonable price), thus increasing the amount available to members. (2) We have claimed that the cohesion of minorities tends to strengthen during difficult times and that the Karaites show flexibility and the ability to improvise solutions for changing reality. This appears to have occurred in Ranen. The crisis in the agricultural sector led the Ranen's members (at the end of 1987) to elect a more cohesive and stronger Executive Committee to guard the financial system from collapse. To be able to meet this goal, the Committee recruited Pinchas Gal based on the experience the moshav had with him. Damages from very cold weather in the winter of 1989 resulted in a financial crisis (difficulties in paying debts) and there was a need to devise a means to alleviate the crisis. An even stronger Executive Committee was elected to resolve the financial crisis. It started to negotiate a debt rescheduling agreement within the framework offered by the government to alleviate agricultural crises in general. (3) The fast increase in the production of tomatoes and flowers in greenhouses, much of which was marketed by the cooperative institutions (i.e., a rise in the share of collectively marketed produce) resulted in an increase of the Association's cash flow and a further strengthening of the Association. (4) The increase in the Association's cash flow further increased the strength of its institutions since members' trust in them improved.

These four factors indicate a gradual strengthening of the Association. Moreover, they have accelerated the return to agriculture, and the return to agriculture further strengthened the Association. This is true only if the return to agriculture increases the collectively marketed produce and, in turn, the Association's cash flow. In fact, the strength of the Association and the return to agriculture are the cause and effect of each other.

Furthermore, members who had deserted agricultural activity in favor of other nonagricultural occupations started investing in greenhouses (for tomatoes and flowers) to produce for export. Others who grew field crops and vegetables for the local market also joined the process of investing in capital-intensive greenhouses.

It appears that the initiative to develop greenhouses was the result of a decline in the profitability of open field crops and the drastic cut in the water quotas. The situation called for a solution that would allow both a cut in the size of the irrigated areas and an increase in the profitability per unit of water. Such a solution was embodied in the greenhouse agriculture. Following talks with the settlers, one is led to believe that this was indeed the motivation which led to this change in the structure of the farm units.

Table 11.1 summarizes the characteristics of agriculture in Ranen. The turning point is 1988, when the decline in the number of full-time agriculturists stops and the trend is reversed. In addition, greenhouse area increases impressively. The data show a move from extensive agriculture (orchards, open field vegetables, and the like) to a combination with intensive greenhouse agriculture parallel with the drastic cut of the water quota within several years. Greenhouse activity, the growing of tomatoes and flowers, includes about fifty farm units.

According to the members, most of the revenue from the sale of agricultural produce is handled and managed by the Executive Committee of the Association. Therefore, the annual revenue data as recorded in the Association's books are a good indicator of the total financial activities of the members. Table 11.2 shows the decline in revenues during the period of 1986/88, followed by an increase in 1989 back to the real level of 1986.

**Table 11.1**
**Agricultural Activities**

| Characteristics | | 1987 | 1988 | 1989 | 1990 | 1991 |
|---|---|---|---|---|---|---|
| No. of populated units | | 80 | 80 | 80 | 80 | 80 |
| Occupational Distribution | | | | | | |
| Work on the Farm Only | | 30 | 23 | 30 | 36 | 45 |
| Work Outside & Farm | | 40 | 52 | 45 | 40 | 29 |
| Work Outside the Moshav | | 7 | 2 | 2 | 2 | 6 |
| Agriculture: Area (dunam) | | | | | | |
| Orchards (w/o citrus) | | 1011 | 1085 | 1173 | 1173 | 650 |
| Greenhouses: | Flowers | 4 | 5 | 40 | 100 | 134 |
| | Tomatoes | | 23 | 50 | 76 | 129 |
| Vegetables (open field) | | 835 | 805 | 765 | 965 | |
| Ann. Water (Thous. c.m.) | | 1797 | 1550 | 1374 | 1316 | 1016 |
| Poultry: | Quota (in Tons) | 477 | 477 | 477 | 477 | 477 |
| | Actual Production | 355 | 459 | 477 | 471 | |

**Table 11.2**
**Revenues of Collectively Marketed Produce (in Thousands Shekels)**

| Branch | 1986 | 1987 | 1988 | 1989 | 1990 | 1991 |
|---|---|---|---|---|---|---|
| Eggs | 47,389 | 43,904 | 63,079 | 108,876 | 84,295 | 88,000 |
| Poultry | 743,706 | 632,066 | 659,711 | 1040,719 | 1226,468 | 950,000 |
| Fruits | 821,661 | 1378,653 | 1114,300 | 1004,163 | 725,840 | 700,000 |
| Potato | | | | | 226,373 | 226,000 |
| Onion | | | | 337,078 | 176,700 | 176,000 |
| Celery | | | | | 49,871 | 146,554 |
| Pepper | | | | | 115,157 | 103,357 |
| Other Vegetables | | | | 208,285 | 180,292 | 184,000 |
| Tomato Export | | | | 327,273 | 955,238 | 1,350,000 |
| TOTAL VEGETABLES | 541,920 | 176,874 | 242,299 | 1,037,664 | 1,788,514 | 1,936,000 |
| Flowers | 67,440 | 84,936 | 102,200 | 747,246 | 1,100,700 | 1,300,000 |
| Citrus | 266,455 | 293,682 | 217,766 | 264,283 | 408,275 | 250,000 |
| Field Crops | 194,469 | 283,654 | 46,903 | 121,447 | 163,544 | 170,000 |
| ANNUAL TOTAL | 2,683,040 | 2,893,769 | 2,446,258 | 4,324,398 | 5,497,636 | 5,394,000 |
| ANNUAL TOTAL IN 1991 VALUES | 6,076,36 | 5,643,575 | 4,098,762 | 6,003,384 | 6,489,269 | 6,116,318 |
| $ (year end) VALUES | 1,805,059 | 1,880,781 | 1,451,785 | 2,202,954 | 2,684,393 | 2,679,318 |

Comments: 1. The data are from the Association's annual financial report, all in nominal values. 2. The last row of calculations assumes end-of-year values.

The data presented in the Tables 11.1 and 11.2 demonstrate what one of the members pointed out as the characteristics of the Association. The first one, and possibly the most important, is the existence of a significant collective (organized) marketing of the produce. As Pinchas Gal said, "The major part of the produce is marketed collectively. The ability of the Association to offer assistance, like in the liquidation of the sheep, egg and carnation branches, depends on that collective marketing, which alone can provide the Association with the necessary cash flow."

In this context, we should point out that twenty years ago when the sheep ranches were set up, the General Assembly of the moshav had already resolved that for every productive project, the proportion of self-financing must attain at least 50 percent. Even if appropriate loans could be obtained from outside sources, the project would not be approved without the required self-financing.

An additional cluster of characteristics is composed of courtesy, mutual respect, thrift, and the emphasis on self-labor. Although this last requirement was relaxed through the years, it was emphasized that the owner of the farm unit should at least work alongside his salaried workers. These characteristics alleviated social tensions and prevented confrontations caused by the scarcity of agricultural means of production and disagreements regarding their allocation by the Executive Committee. They added to the ability of the Committee to resolve issues without having recourse to outside assistance.

A reason for this might be that Karaism emphasizes the prohibition of evasion and concealment in commercial (or any other) negotiation (Dvir 1990). Although this prohibition prevails in Rabbinic Judaism as well, its influence seems more significant in the small, intimate Karaite community, especially in a moshav like Ranen. All of these characteristics together make the important third characteristic, namely, a strong Association. As summed up by one of the members, "A strong Association saves bureaucracy, instills confidence, and makes it easier to procure credit. It also helps the membership to weather crises."

The strength of communal ties is shown by the fact that confrontations that occurred from time to time were almost always resolved locally by means of an arbitration committee or a single arbitrator elected from among the members. Only in one case in Ranen's forty years did confrontation lead to external arbitration and eventually to trial in court. The conflict hails back to the early 1980s, when many of the poultry farmers stopped producing their quotas due to a decline in profitability. Four poultry farmers continued to produce the whole quota of Ranen. In 1986, when profitability improved, several poultry farms returned to operation, and the Executive Committee reinstated the original distribution of the quotas. The owners of the four poultry farms turned to the court since they felt that the Executive Committee had misinterpreted the arbitrator's decision. The court resolved the issue by restoring the equal quota system with the exception of two members who received twice the quota of others.

The desire to strengthen the Association was an important reason for the business initiatives that led to the impressive economic development in 1988. Economic and financial activity of the Association in the winter, as with every farm in which vegetables and fruit are grown, was small. The effort to develop winter greenhouse agriculture was aimed at diversification to reduce the risks incurred by the farmers and to spread the Association's cash flow over the entire year. It also created a new source of income for the poultry growers and made it possible for those who lost their jobs outside the moshav to return to agriculture. Finally, the new branch increased the size of the resources to be divided among the membership and reduced the tensions due to disagreements over the distribution of production quotas.

The success in Ranen—the secret of survival, as one member said—is derived from stubbornness and perseverance even in time of crisis. The reason for this, he pointed out, is the close relationships and personal economic ties already existing from their life in Egypt when they worked in commerce and gold/silversmithing and lived together in their own neighborhoods in Cairo. In short, it seems that the fact that the moshav is populated by Karaites alone strengthens the community of Ranen.

## CONCLUSION

Since the mid-1980s, the Association in Moshav Ranen has been growing stronger, and its members have been returning to agriculture, thus going in the direction opposite to that of most of Israel's moshavim where Associations have been falling apart. We find that the two processes are related.

Furthermore, without the strength and strengthening of the Association, it would have been harder for the member to raise the capital necessary for a return to agriculture, especially in capital-intensive branches. And without the turnover made possible through collective marketing by intensive agricultural activities, the Association could not have been as strong as it was. We believe that there is a connection between Ranen's Karaite religious affiliation and important patterns of behavior. These are mutual respect, trust, assistance, personal responsibility, social cohesion, self-employment, and frugality. These characteristics, essential elements for successful operation of any cooperative association, are reinforced in Ranen by customs of the Karaite community, such as decision making by consensus. The entrepreneurial factor usually exhibited by religious minorities and the increased cohesion that they tend to show in times of crises are also contributing factors.

All of these factors, together with the method of collective marketing, alleviated crises, prevented confrontations, and strengthened the Association in Moshav Ranen. Furthermore, the collective marketing methods enabled the Executive Committee to effect the transition to more capital-intensive agriculture (i.e., greenhouses), which further strengthened the Association and improved its prestige.

The economic data presented here show that the decrease in agricultural activities from 1986 continued until 1988. In 1989, the real revenue level of 1986 was restored. Similar processes of recovery had already occurred several times in Ranen in its short history. While one should be careful in generalizing from a single case, one could say that safeguards against cheating, such as the rule of self-financing (at least 50% of any investment), are vital if one is to ensure the survival of a strong association. Proper leadership of the moshav and personal responsibility of the individual agriculturist is another important combination of characteristics. Apparently, what happens to social norms and communal rules of operation is not less important than economic factors if one is to properly understand the collapse and disintegration of the moshav associations in Israel.

## REFERENCES

Algamil, Y. B. O. 1981. *The Karaite Jewish History*. Ramla: The National Council of Karaite Jews in Israel.

———. 1985. *Karaite Jewry in Egypt in Modern Times*. Ramla: The National Council of Karaite Jews in Israel.

Ankori, Z. 1959. *Karaites in Byzantium, The Formative Years, 970–1100.* New York: Columbia University Press.

Applebaum, L., D. Shoresh, M. Eger, and S. Hartman. 1990. Alternatives to Association in the Moshav. Research report. Rehovot: Center for Development Studies.

Colligan, S. E. 1980. "Religion, Nationalism, and Ethnicity in Israel: The Case of the Karaite Jews." Ph.D. diss. Princeton: Princeton University, Department of Anthropology.

Corinaldi, M. 1984. *The Personal Status of the Karaites*. Tel Aviv: The Faculty of Law, University of Tel Aviv.

Dvir, Y. 1990. "Information Pamphlet on the Karaite Jews." Ramla, Israel: Keter.

*Encyclopaedia Judaica*. 1972. Vol. 1. New York: Macmillan.

Haruvi, N., A. Fleischer, and M. Eger. 1989. "The Employment Structure in the Moshav Sector." Rehovot: Center for Development Studies.

Kislev, Y., Z. Lerman, and P. Zussman. 1989. "Credit in the Moshav—Experience and Lessons." *Economic Quarterly* 40(141): 114–131 (Hebrew).

Mahler, R. 1949. *The Karaites, A Mediaeval Jewish Movement for Deliverance*. Merchavia: Sifriat Poalim.

Nemoy, L. 1952. *Karaite Anthology, Excerpts from the Early Literature*. New Haven: Yale University Press.

Weber, M. 1958. *The Protestant Ethics and the Spirit of Capitalism*. New York: Scribner.

PART III

# FROM GENERATION TO GENERATION AND FROM COUNTRY TO COUNTRY

Chapter 12

# The Moshav Overseas: Revisiting Issues Rather Than Settlements

*Yair Levi*

*While the moshav was shaped to address Israeli's own objectives, by the 1960s, the moshav model was being promoted as an appropriate, even ideal, solution for the organizational needs of rural development in the Third World. Yair Levi presents the reasoning behind this conviction not only by Israeli development experts but by Third World adherents. Levi then goes on to review the outcomes in a series of cases in Asia, Africa, and Latin America and to analyze reasons for their divergence from the Israeli experience. While specific historical circumstances influenced the progression of events in each case, some general themes emerge that could offer correctives to present and future initiatives in the transfer of both technology and organization in the process of overseas assistance.*

*The experience with attempts to export the moshav model to Third World countries was a relatively short-lived one. Started in the early 1960s, it was followed in the mid-1970s by a withdrawal of most of the Israeli teams due either to operational considerations or to the political events of 1973. This stage in the life of the settlements was not followed by a devolution of control to the settlers as foreseen by the original model but rather by the discontinuation of most projects. This chapter analyzes the early motives and beliefs that justified the experimentation with moshav settlements overseas, reviews a few cases in Asia, Africa, and Latin America and revisits the issue in the light of current development views and trends.*

Revisiting the experimentation with the moshav in Third World countries is a stimulating experience for two reasons. First, it enables one to analyze a most interesting chapter in the Israeli assistance to overseas countries and to point to the differences and the similarities in moshav policies inside and outside Israel. Second, and most important, such a study permits an overview of the early motives and ideologies that justified the export of the moshav in light of the main changes that occurred throughout the past decades. These include such basic developmental issues as the "top-down versus bottom up" argument in development policies; the pitfalls of quick technology transfer; the place of local traditions and patterns in rural development; and the issue of government-to-government versus movement-to-movement technical assistance.

This chapter argues that it was wrong to assume that the moshav model would suit the intended beneficiary groups on the account of being more accessible than the kibbutz and the *moshav shitufi* (where domestic life is household based as a moshav ovdim but economic enterprises are communal as in a kibbutz). In this respect, the overseas experience showed great similarity with the domestic experience in Israel in the attempts at quick socialization of new immigrants in the 1950s.

This point is discussed in a paper (Levi, in press) that focuses on the domestic experience of the new immigrants' moshav (moshav olim) of Israel. It is shown that the decision to settle the new immigrants in moshav-type settlements was based on the comparison between the three historical forms of Israeli settlements: the kibbutz, the moshav shitufi, and the moshav ovdim. The latter was seen by the planners as the least challenging in terms of the scope of activities and the level of ideological commitment demanded of the members. Its family-based farms, it was assumed, would make management relatively easy. As such, the moshav was seen as the one better fitting the culture of immigrants who came to Israel from traditional Third World settings.

In this chapter, I argue that these development assumptions were challenged by the actual experiences of settlers. Type of management did not correspond in a predictable way with the two other dimensions (scope of activities and ideological commitment). Thus, the decision to implement the moshav ovdim model was based on fallacious reasoning on two accounts: first, because the decision about what is better was taken by the planning authorities only and, second, because the comparison between high and low levels of cooperative organization failed to take into account other forms (apart from the three main settlement types) that could be found as early as the 1950s in both Israel and overseas countries. This point is further elaborated on in the conclusion of this chapter.

Although at odds with previous studies on this topic, this chapter will suggest that there were two chief factors responsible for the poor performance of moshav projects in Third World countries. The first was their top down origin. Second, these model villages failed to meet the basic criteria for areawide socioeconomic compatibility and replicability. In what follows, we shall review the circumstances and the belief system underpinning the attempts to transfer the moshav model to Third World countries.

## THE MOSHAV-TYPE SETTLEMENT IN THIRD WORLD COUNTRIES

### General Background

A number of circumstances in both Israel and abroad contributed to the export of the moshav model overseas starting in the early 1960s. Many countries in Africa, Asia, and Latin America, perceiving a rural development problem for the first time, were eager to learn from a young country where cooperatives had played such a central role in nation-building. By the early 1950s, international agencies were already impressed by the spectacular increases in food production achieved by the moshav sector within the span of a few years. It was particularly important that these results were obtained by people who were generally without prior agricultural experience and who came from Third World countries. This created the impression that Israel had found the key to quick agricultural development.

In Africa, many policymakers and thinkers viewed full-fledged village cooperatives as a tool for the quick Africanization of the rural sector. This entailed the elimination of abusive middlemen and private money lenders, most of them of foreign origin. Moreover, cooperatives were seen as a valid tool in the development of a socialist system. To the Nigerian writer Onouha (1965), "African socialism has every reason to espouse the growth of cooperatives, since it corresponds so perfectly to our traditional method of ownership . . . and it fosters the community spirit . . . agriculture should become our cooperative sector par excellence."

African leaders such as Julius Nyerere (1962), Leopold Senghor (1961), and Kenneth Kaunda (1967) viewed the villagization of rural areas through cooperatives under strong state guidance as a mainstay of their socialist way to development and as an alternative to capitalism and class struggle, both alien to African culture.

Against such a background, the Israeli experience with village cooperatives could not fail to exert its appeal. To African leaders who visited Israel, the rural cooperative success appeared to be intimately associated with political independence. What had preceded Israeli statehood remained for most of them too remote to be understood. Furthermore, they ignored the vital role played in Israel by the supporting service organizations, such as extension, marketing and supply.

The following words by Phillipe Tsiranana (1962), then President of Madagascar, can well depict the euphoric climate of those days:

> I ask all leaders of Madagascar to be at the head of a cooperative . . . [in Israel] all Ministers and Members of Parliament lead a kibbutz or moshav . . . we have to imitate this example if we are to be true patriots . . . we have 107 deputies, 54 senators, and several hundreds of generals and municipal counselors. If each of them agrees to lead only one cooperative, the entire island would be soon covered by cooperatives.

This positive attitude toward cooperatives as tools of national development failed to acknowledge warnings about the dubious relevance of African communitarian habits to modern socialism (Meister 1966) and the poor chance that Western cooperatives could be successfully grafted upon traditional milieu (Texier 1969). As for the more general problems inherent in state initiated and led cooperatives as a tool of rural development, three decades of failures were still ahead, so the time was premature for those leaders to be aware of the risks involved.

This infatuation with the cooperative came at a time when in Israel the cooperative movement was already showing symptoms of crisis. The new cooperative bill proposed by the Moshav Movement in the 1960s included regulations aimed at defending the moshav from its own ideological erosion. Through assistance to developing countries, rural experts from Israel, many of them members of moshavim, found a new outlet for their energies and an opportunity to share their rich experience with others. Thus, domestic problems acted as incentives to export their knowledge thus enabling them to become agents of overseas change agents in rural—mainly moshav—ventures.

Although settlements constituted only a part of the overseas projects using Israeli expertise, it was precisely the newness of the

Israeli settlement patterns that attracted leaders from developing countries in their search for new avenues of development. It was hoped that by emulating the Israeli model cooperative villages would serve a number of national aims. These included resettlement schemes wherein networks of villages would absorb ex-army personnel, school-leavers, refugees, landless people, and so forth (Nigeria, Nepal, Iran); the establishment of new isolated villages in areas of particular importance for peasant organization and the demonstration of new farming methods (Zambia, Dahomey, Thailand, the Dominican Republic); or the establishment of pilot centers with a view to training youth in the framework of agricultural pioneer movements (Togo, Central African Republic).

## THE JUSTIFICATION OF THE MOSHAV MODEL

The introduction of the moshav-type model in Third World countries was not a straightforward process. It was accompanied by much discussion in development circles as to whether settlement projects were preferable to reforms in traditional milieu. Advocates of the first approach argued for the potential demonstration effect of model villages on surrounding areas, thus justifying high initial investments per farm unit. Others expressed a preference for reform in existing villages, which apart from lower costs was based on several factors. These included the need to avoid the emergence of a new gap between privileged settlers and the average farmer; the need to reach as many rural beneficiaries as possible; and a wish to avoid a drastic break with subsistence farming and local traditions. This controversy varied according to local conditions and policies. In Nigeria, the settlement approach inspired by Israel in the early 1960s was warmly supported both by Nigerian scholars (Oluwasanmi 1965) and then Minister of Agriculture (Kreinin 1969) in sharp contrast with experts of the Food and Agriculture Organization (Kreinin 1964). In Kenya, the advice of Israeli planners and local authorities was rejected with respect to the development of the Yatta Valley in favor of an areawide and a more restricted, selective strategy. In some Francophone African countries, the Israelis attempted to combine the advantages of both systems by using the traditional village as a point of departure for the creation of new cooperative villages in its proximity. Alternatively, new agricultural centers were created to train farmers from the area in modern techniques to be introduced at a later stage in their original villages. In some cases, a combination of military training with settlement of young people was adopted after the Israeli model of the Nahal.

Whatever the case, however, the basic principles underlying the Israeli approach remained largely those of the moshav ovdim model: the all-settler cooperative with the family as the basic unit of production without dependence on hired labor; the indivisibility of the family farm; and the aim at equalizing the farmer's income to that of the average urban worker, all these under external guided planning and assistance. This was the approach of Moshe Dayan, then Minister of Agriculture (1964), with regard to the Yatta Valley project in Kenya. In his evaluation of the moshav replicability, Desroche believed that three of its main traits—that is, (1) nationally owned land leased to the farmer; (2) the coterminous multipurpose cooperative and residential unit; and (3) the combination of the cooperative system based on proportionality of contributions and mutuality based on helping the needy—would enable the moshav key to fit into locks of traditional villages characterized by the persistence of community practices (1973). All in all, the moshav-based approach was seen as the distinctive mark of the Israeli assistance as compared to that of other countries. Whereas in the latter emphasis is put on the provision of services to farmers, in the first what predominates is the holistic interpretation of rural assistance as a tool aimed at radical changes in the peasant's life (Levi 1970). Israeli representatives abroad displayed a clear pride in pointing to this distinctiveness. In a report to Jerusalem dated February 28, 1966, the Israeli ambassador to Cameroon refuted the content of an article (*Le Monde* 10/2/1966) criticizing the Israeli activities at Obala (one of the centers for model extension and new moshav-type settlements) by saying that "while the French are content with training centers, we have moved a step forward to the creation of new villages" (Israel's Ministry of Agriculture, File Cameroon).

The justification of the moshav model was broadly similar to that adopted in Israel during the 1950s, a period of high immigration. As mentioned above, the moshav was seen as the lowest of the three main patterns of cooperative settlements in Israel (kibbutz, moshav shitufi, and moshav ovdim) and consequently the most suitable for absorbing people from rural traditional settings. Commenting on the moshav adaptability to Africa, Frank (1968) made the point that in spite of its limitations, "the moshav ovdim formula in general and its application in the form of immigrant moshav in particular, seems to be the most congenial pattern of Israeli cooperative land settlement for adoption by the developing countries of Africa." However different the backgrounds of the new immigrants in Israel and the African peasantry, "This type of settlement can offer a favorable pattern of agricultural organization to people who are not acquainted with a modern agrarian economy. In contrast to the kibbutz or the moshav shitufi, whose complex large-scale administration is still a handicap under present conditions in Africa, the moshav ovdim formula has a fair chance of succeeding gradually."

In a more general way, Schickele as quoted by Kreinin (1963) considered "the Gezira scheme in the Sudan and the Israeli moshav as the systems offering the best solutions to the organization of agricultural production in Africa." In his in-depth study of African rural traditional societies, the French sociologist Gosselin (1970) spelled out an elaborate justification for the adoption in the mid-1960s of a moshav ovdim type of orientation by then Dahomeyan Société Nationale pour le Développement Rural (SONADER). The latter was set up in 1962 as a public institution for rural development through the creation of cooperatives to enhance the palm oil sector and secure its development and sustainability. Gosselin was confident in the success of the SONADER cooperatives on the ground that it was a promising attempt to rationalize production through the integration in the same system of landowners and landless peasants. Gosselin's optimism was based on his theory that modern cooperatives would succeed to the extent that they could substitute for already disintegrating local systems. Such was the case with the *donkpé*, a traditional form of collective work at the village level in southern Dahomey.[1] A major argument in Gosselin's favorable diagnosis of the chances of the new cooperatives was the SONADER pragmatism:

> The SONADER cooperatives reject the state controlled *sovhoze* as well as the collective kibbutz. They are closer to the *kolhoze* sector or the *moshav ovdim* which leaves to the peasant the managerial autonomy of his farm or part of it, yet obliges him to abide by indispensable rules of discipline and to integrate into a state or cooperative sector for basic investments and farm operations, total or partial marketing or products etc.

The introduction of the moshav in Dahomey has to be seen in light of the overall infatuation for the Israeli settlement models that pervaded the country at that time. Frank (1968) has illustrated this point with two cases: First, the Doja project, initiated in 1959, settled 100 families on 150 hectares for the cultivation of palm trees and other crops. According to Frank, "One can feel the desire to imitate the Israeli kibbutz or moshav, whose example seem to have impressed the people of Dahomey . . . the idea is to assemble the peasants on a given place where intercalary crops will be grown until the palm trees become productive . . . the attempt is to create totally new communities." Second, the Houin village project aimed to cultivate palm trees on an area of 4,000 hectares. Here, "out of the village's 3,000 inhabitants, about 200 have already joined the new organization, which is inspired by the Israeli *moshav ovdim.*" It appears that the preference for the moshav by foreign observers was not always based on a sound knowledge of the same. On the whole, the point has to be made that Israel's planners and consultants never tried to introduce a collective system like the kibbutz or the moshav shitufi in projects carried out in developing countries. The attractiveness of the moshav ovdim model stemmed from the fact that here the family constituted a unit of both consumption and production. Hence, it was simply assumed to be suitable for traditional family-based settings.

## AN EVALUATION OF MOSHAV EXPERIENCES

### General Considerations

In his overview of experimentation with moshav-type settlements in developing countries, Kahan (1986) listed fourteen of the most known projects, of which seven (50%) were in Africa, five (35.7%) in Asia, and two (14.3%) in Latin America. These were carried out from the early 1960s on. The general conclusion was that in spite of the apparent a priori advantages of the moshav formula, its application to the developing world was in many instances unsuccessful: "The central issue was one of the degree of transferability of one type of experience to alternative environments. . . . The experience of organizational transplantation in most situations is likely to be doomed from the outset," hence his call for more flexibility in institutional planning and implementation, varying not only over space but also over time. In Kahan's (1986) opinion, changes over time in the economic structure of cooperative villages may result in the following policy alternatives:

1. Omitting the municipal activity of the cooperative.
2. Restoring or eliminating some of the social or economic goals.
3. Dividing the scope of services between primary and secondary cooperatives.
4. Reforming the cooperative on a homogeneous farm enterprise basis.
5. Dissolving the village cooperative for second level, regional, or national cooperatives.

These policy alternatives touch upon three basic dimensions, namely, the scope of activity, the extent of heterogeneity/homogeneity of the farm enterprise, and horizontal–vertical integration.

Issues of organizational and cultural compatibility to the local fabric are mentioned in Kahan's analysis yet are not presented as key variables for rural planners.

Our argument is that virtually all the moshav projects failed to meet basic criteria of weaning, conceived as the state of maturity needed for a planned settlement to become self-sustaining and based on settler organizations. Though not always explicitly included in the official moshav technological package aimed at overseas countries, the implicit rationale of the moshav, and for that matter, of any cooperative organization, implies that sooner or later the cooperative will be managed by the members themselves. This objective was clearly aimed at in the moshav ovdim model (which inspired its export to overseas countries) and was attempted through a sophisticated and gradual precooperative policy by the settling authorities in Israel.

In our study of self-management in new land settlement projects (Levi and Naveh 1988) "weaning" was operationalized by means of four main components, two from "above" and two from "below." The first consists of the settling authorities' readiness for devolution of management of the settlement to the settlers and the availability of a support system enabling the settlement to function. The second consists of the extent of acceptance by the settlers of the organizational system of the settlement and their overall maturity for common action. As a composite measure of maturity from above and from below, full weaning can result only from the combined effect of the four components in turn affected by a variety of independent variables not to be examined here.

To evaluate the extent to which moshav experiences managed to meet the above weaning criteria, as necessary though insufficient conditions of self-sustained development, six cases will be examined. Two (Argentine and the Dominican Republic) are taken from our above study (Levi and Naveh 1988 and our field work in 1981), two (from Nepal and Thailand) are based on the author's personal visits in 1984, and two (Dahomey and Nigeria) are based on written accounts from the extant literature. Though not exhaustive of the whole universe under consideration, these cases, it is believed, present a typical selection of experiences based on single and multivillage moshav projects.

## An Overview of Case Studies

*Argentina.* A plan for the integrated development of the Catamarca Valley in northern Argentina was started in 1965. An Israeli team participated in the elaboration of a three-settlement project whose physical layout and farm structure were modeled after the moshav type. Of the three settlements, only one—Nueva Coneta—became operational in August 1969; the others did not reach beyond the planning stage. The first settlers, of Spanish and Italian origin, arrived from the neighboring areas where they used to be landless workers and sharecroppers, well-experienced in grape and onion growing. Upon their arrival, they received houses and twenty hectares of land, which included four hectares around the house and sixteen in farther plots. Membership in the settlement cooperative grew from 25 to 136 in 1973, yet decreased to 51 by the end of 1981. This decline was due to strict application by the village cooperative of the statutes with respect to capitalization and debt repayment: deviant members were expelled. The initial full support by the Provincial and Regional Authorities was reduced in time to the delivery of municipal services and the control of physical infrastructures. The Cooperative managed to buy a factory for dehydrating vegetables and to meet the onerous conditions of debt repayment. At the time of our study (summer 1981), about half of the 120 settlers of Nueva Coneta were not members of the original all-settlement cooperative. Many had to attend to their own production, credit, and marketing operations. Others were members of new organizations that emerged in the meantime in the settlement, such as a joint venture for the upkeep of a nursery of citrus trees and a cooperative of dairy farmers. Nueva Coneta seemed to operate on a self-managed basis yet as a microcosm in a nonsupportive environment. The internal structure was one of plurality of organizations, at odds with the original moshav inspired model. The growing tendency by relatively well-off people from outside to buy farms in the settlement and turn into absentee-owners through the employment of hired workers signaled a clear departure from the original objective to improve the situation of local landless peasants by turning them into cooperative settlers. Nueva Coneta may illustrate an instance of partial weaning characterized by maturity from below and devolution from above yet without the availability of a support system.

*The Dominican Republic.* Haras Nacionales, about twenty kilometers from the capital Santo Domingo, was founded in 1973 by the IAD (Instituto Agrario Dominicano), the government body in charge of land reform. From the start, the settlement was under the special auspices of the president of the Republic with the aim of eventually becoming a pilot dairy project shaped according to a moshav village. By 1975, about fifty-six families had been settled. Farming was based mainly on dairy and a few field crops. Each farmer was due to receive about ten cows and 150 tareas of land (1 tarea = 0.628 square meter).

An attempt to create a settlers' organization in 1975/76 failed due to internal fighting and external interference. This event left the prolonged tutelage of the IAD practically unchanged, along with Israeli assistance in dairy farm management. An evaluation made in 1980 by the then Israeli expert in Haras Nacionales showed that milk production had risen from a total of 13,700 libras daily in 1978 (1 libra = 450 centiliters) to 35,000 libras. An economic plan included the increase of cultivated area for traditional crops and the introduction of new ones. First priority was to be given to the devolution of management to the settlers (Zafrir 1980). At that time, decentralization of the administrative structure and the issue of self-management were widely discussed by a seminar organized by the IAD. It was felt that the settlements of IAD should strive for self-management of the settlers "in such a way that the beneficiaries participate from the beginning in the management of the enterprise, because the capability for self-management is undermined if the dependence of the settling authority, in this case IAD, is maintained . . . this is very important because the process of our agrarian reform has taught us another type of administration" (Gil and Oliart 1980).

We lack information as to the present situation. At the time of our study, despite declarations of a new policy, Haras Nacionales was in its twelfth year of external administrative management. The emerging picture was thus one of nonweaning combined with relative economic success.

*Nepal.* The two projects with Israeli assistance—Nawalpur and Nepalgunj—are among the widest ever attempted by Israel in developing countries. The original planning of the Nawalpur project (the one of our concern here) provided for the settlement of landless people from the Terai area of Nepal was soon increased by hundreds of Nepalese refugees who arrived from Burma in 1964. The idea was to create clusters of three cooperative villages of sixty to one hundred family units each located around a common service center. The detailed layout provided for one bigha (0.70 hectare) for each settler's homestead and a plot to be cultivated for house consumption and separated contiguous blocks of three bighas to each family allowing for mechanized farming and efficient pest and disease control. The physical planning was broadly modeled after the Israeli multineighborhood settlement (Weitz and Rokach 1968). The cooperative plan was a replica of the moshav frame. Every settler was to be organized in a cooperative society in charge of all farm input and output operations under strict supervision. Every family was to receive the same amount of land and to be responsible to cultivate its own land under a system of supervised credit controlled by the cooperative (Reiss 1969). A central cooperative was to comprise five subcooperatives with one cluster of villages each and to operate a mill for husking rice, grinding flour, and extracting oil. The whole cooperative system was put under the supervision of PUNARVAS (the Nepal Resettlement Company), a government body established upon Israeli advice in 1964 (Prives 1977).

By 1971, the Nawalpur project comprised 1,200 families, about 8,000 people, cultivating 8,843 bighas, of which 2,436 were maize, 2,280 were unirrigated mustard, 132 were rain-fed wheat, and 3,995 were for overall cultivation. The yearly income of an average family totaled 6,500 NR (in 1971, US$1 was equivalent to about ten Nepalese rupees) which compared most favorably with the then officially published national average of 2,000 NR. The villages were served by seven elementary and two high schools, and the opinion prevailing among the Israeli team experts was that all this was auguring well for an imminent weaning from foreign assistance.

The Israeli expert withdrew from the Nawalpur and Nepalgunj projects at the end of 1975, "having transferred the maintenance of two well-established settlement areas to the personnel of the PUNARVAS Company" (Prives 1971).

In 1984, when we visited the area, we were told that PUNARVAS had withdrawn by granting land titles to the settlers without setting regulations as to land policy and this had caused widespread fragmentation, sales of land to new settlers, demographic pressure, and wide ethnic heterogeneity, in turn causing many original settlers to move out and to return to traditional crops. Nothing was left of the erstwhile cooperative structure, and farmers were obtaining credit and marketing on their own. The rejection of the planned cooperative system was attributed to its inadaptability to local conditions and the inadequacy of PUNARVAS's policies to cope with the complexities of settlement withdrawal beyond the formalities of granting land titles. The new policies of PUNARVAS in the Nawalparasi area did not show departures from the early top-down approach and the absence of a community-based and participatory orientation. The pattern of Nawalpur thus suggests that basic preconditions for successful weaning were missing.

*Thailand.* Situated about forty kilometers from the town of Patchabury in southern Thailand, the settlement of Hubkapong originated from the initiative of the king to create a model village with Israeli assistance. The Thai–Israeli development project was approved in 1966. In addition to the donation of 12,500 rai of land, about 130 families were settled and each given twenty-five rai of land (2.5 rai = 1 acre), of which seven were for home plots partly irrigated and the remaining eighteen depending solely on rainfall. From an initial status of precooperative, the village was registered as the Hubkapong Agricultural Cooperative in 1971 with the withdrawal of the Israeli team. In spite of scarcity of water for home consumption and farming, the village displayed a highly-developed farm system at the time of our visit in 1984. The main crops were asparagus, supplied to Thai International and Singapore Airlines, sugar, sisal, and peas. As a supplementary occupation, the promotion of handicrafts was helped by the personal involvement of the queen in the sales of products in Bangkok. From 1972 to 1981, the cooperative membership had increased from 135 to 294 (+117%), the share capital from 253,647 to 5,218,113 baths (+1,957%), the reserve fund from 60,661 to 544,667 baths (+797%), and the volume of sales from 118,327 to 3,468,343 baths (+2,831%) (Hubkapong Cooperative Village 1982). In 1984, the 300 cooperative members were electing a management committee of nine (of whom three were changed every year), in turn appointing a manager, a secretary, and a treasurer. The manager of the project would participate in the meetings of the committees and enjoy the right to veto whenever a decision contradicted the statutes. The achievements of Hubkapong have led to the creation of eight similar model villages in other parts of Thailand, all under the patronage of the king. The pattern of Hubkapong is one of economic success under a formal cooperative system, still devoid of participatory settler organization (i.e., maturity from below).

*Dahomey.* The aforementioned experience of the SONADER was fraught with failures. The cooperatives did not exactly operate according to the moshav model, as they were based on obligatory pooling of land (though maintaining individual land rights) and envisaged an unconventional combination of nonworker-members, owner-members, and nonowner workers eligible for membership through acquisition of a share of 200 working days. By the end of 1966, SONADER was operating on an area of 7,000 hectares with more than 7,000 members with different types of shares (International Labor Organization 1967, quoted by Gosselin 1970). Seen in retrospect, the diagnosis and optimism of Gosselin (1970), as earlier mentioned, seems totally unwarranted.

This can be illustrated by the Benin Hinvi Project for palm oil cultivation approved by the World Bank in 1990 at a cost of US$9.6 million to benefit some 4,000 farmers. The project transformed about 75 percent of the land of the area into blocks for palm oil production and allocated 25 percent for the cultivation of high-yielding crops. The project failed to take stock of the fact that the 75 percent land was traditionally used by women for palm oil production, for gathering wood for fuel, and left fallow to preserve soil fertility. The compulsory cooperative of the SONADER and the refusal of the Project Administration to raise prices of palm oil to the producers caused the latter to neglect the main project's activity. Farmers gave priority to their food needs, yet now on one-quarter of their previous land. Without rotation, the soil deteriorated and yields began to drop. By 1983, the area had turned from a food surplus to a food deficit one. The women had lost their income sources and what was left of the traditional life and production patterns were destroyed. The case of Dohomey is reported as one of the most blatant instances of failure among World Bank–assisted projects (World Bank 1985).

*Nigeria.* Moshav-type settlements were attempted in 1959 and 1962 in the western and eastern regions of the country, respectively. In the first case, the project mainly addressed sixteen- to eighteen-year-old school leavers. Guided involuntary settlement and lack of respect for local culture characterized the 1959 experimentation. In his evaluation of seven years of operation, Eicher (1969) made the point that "moshav farm settlements are, in practice, state farms rather than cooperative land settlements."

To obviate such shortcomings, the project in eastern Nigeria introduced a number of changes, such as settling experienced farmers in the age range of twenty-five to thirty years and an average size of private tree-crop farms (for palm oil and rubber) of seventeen acres instead of thirty as originally established in the western project, as this size was considered too large for a typical farm equipped with machetes and hoes. The lack of experience in multipurpose settlement cooperatives prevailing in the eastern region ministries and the inability of induced settlement schemes to provide increased agricultural outputs in the region led Eicher (1969) to the formulation of alternatives to capital-intensive settlement policies. Evaluating Nigeria's first plan (1962 to 1968) results, the recommendations were that

> the government should reject the moshav cooperative farm settlement approach and rely instead on the imagination and initiative of farmers to construct their own building and develop their own land. . . . Substantially higher rates of return on investment can be achieved by subsidizing a few inputs—oil palm seedlings, fertilizers—for small holders who remain on their farms than from returns to investment in farm settlement schemes. . . . The analysis of the moshav system in Nigeria suggests that the economic history of a nation should be carefully assessed before wholesale importation of a new form of social and economic organization of agricultural production.

After an elapsed time of thirteen years, Oni and Olayeni (1973) assessed the 1959 project as a failure. It had not achieved favorable rates of return on investment, shown demonstrable effects on neighboring farmers, nor alleviated unemployment. The foregoing depicts a situation of total absence of maturity for weaning.

## Analysis

The foregoing review of moshav experiences shows that any attempt at guided settlement under the assistance of a foreign team of experts needs to go through a number of successive stages to pass the test of feasibility and sustainability. These are the stages of weaning from the foreign team assistance, followed in turn by weaning from the control of the national settling authorities and the continuing performance of the settlement under the management of a settler-based organization. Terms such as *feasibility, continuity,* and *sustainability* relate to the much debated issue of what is a successful settlement project. Without entering the complex details of the subject matter, it is suffice to point to two main interpretations. The first views success under a maximalist-transformatory perspective. In the words of Scudder (1981):

> Successful settlements are those that stimulate an ongoing process of integrated area development. Essential to this definition are linkages between rural and urban sectors, with agricultural development stimulating the emergence of a hierarchy of service centers as well as manufacturing and industrial development within the region. The word ongoing means that development must be sustained at least into the second generation.

Similar views about the need for agricultural and industrial integration and for regional planning and integrated area development in connection with settlement schemes, have been expressed by Weitz, Pelley, and Applebaum (1978) and Van Raay and Hilhorst (1981).

Another perspective, a pragmatic and realistic one, lays more emphasis on local needs and on the extent to which as many people as possible can be settled on the land with minimal costs and maximal beneficiary participation (Sawadogo 1979). If the idea is accepted that success is a judgment based on expectations (Ralston, Anderson, and Colson 1983), then a more realistic set of objectives should be set for settlement policies. Common to both views is the importance accorded to settler participation in management, a well-established key success factor in the evaluation literature of settlement policies. Lamenting the paucity of settler participation in fifteen

World Bank–related projects, Scudder (1981) made the point that "a major weakness of settlement projects, this situation tends to perpetuate project management simply because there are inadequate local organizations to take over management responsibilities."

Judged against these criteria of success and our aforementioned weaning, it appears that the reviewed moshav settlements failed to pass the stages of feasibility and sustainability. In no case did the settlement project pass the test of areawide success, neither in its maximalist nor more pragmatic interpretation. In only one case (Nueva Coneta, Argentina) did the settlement achieve a situation of settler management largely autonomous from external control yet deprived of a support system. The case of Hubkapong (Thailand) shows a settlement governed by a local cooperative yet practically dependent on the tutelage of the king. The same two above instances are those where a partial weaning from the settling authorities took place. The first test, the easiest, is weaning from the Israeli technical assistance. Even this was not passed in all the cases, as shown by Haras Nacionales (the Dominican Republic) where an Israeli expert (the third since the beginning) was still in charge of the settlement at the time of our study. Table 12.1 summarizes the foregoing in a schematic way.

Once achieved, the first stage of weaning from the foreign assistance team—a somewhat technical and relatively easy rung on the steep ladder to sustainability—the settlements found it more difficult to achieve even a partial devolution of management by the local

**Table 12.1**
**Development Stages of Moshav Settlement Projects**

| Settlement Project | "Weaning" from the foreign expert team | "Weaning" from the local settling authorities | Achieving settler management | Achieving area-wide effects |
|---|---|---|---|---|
| 1 Nueva Coneta | + | ± | + | - |
| 2 Haras Nacionales | - | - | - | - |
| 3 Nawalpur | + | - | - | - |
| 4 Hubkapong | + | ± | ± | - |
| 5 Benin Hinvi | n.d. | - | - | - |
| 6 West & East Nigeria | + | - | - | - |

authority. This resulted either in the demise of the project (Nepal, Benin, Nigeria) or at best in a continuation of model settlements serving a small elite of rural people (Thailand). The hasty and populist character of moshav policies can be appreciated in the case of Nigeria. Here, the gloomy lesson was that strong cooperatives will take years to develop in Nigeria, and planners who desire instant cooperatives may benefit by a close examination of Nigeria's experience with the moshav (Eicher 1969).

## SUMMARY AND CONCLUSION

The chapter of the moshav export to Third World countries was a relatively short one and could have been evaluated—after its termination—a long time ago. In fact, many Israeli teams withdrew as early as the early 1970s, and others left with the severance of diplomatic relations with African countries in 1973, a fact that possibly precipitated the collapse of projects, as was the case in Zambia. Our view, as clearly ensuing from the foregoing, is that for the majority of the projects a self-sustained development was a priori an inaccessible target.

Revisiting the issue now at a distance of about thirty years enables us, with the benefit of hindsight, to better understand what happened. A kind of could-not-have-been-otherwise stance rests upon the rich experience accumulated in rural development in the past decades. This provides us the following interesting insights.

The first of these is the issue of participation. All the moshav experiences were typically top-down ventures based on government-to-government programs of collaboration[2] and aiming at grafting a foreign model onto totally different settings with little, if any, allowance for adaptation to local idiosyncrasies. Today's literature is replete with the critique of government initiated cooperatives and the call for a rethinking of their role in rural development. Increased reliance on nongovernmental organizations (NGOs), privatization, and even a renewed interest for traditional forms of organization are among the recent alternatives discussed (Verhagen 1984; Uphoff 1987; Braverman, Guasch, Huppi, and Pohlmeier 1991).

A second insight is the issue of the trickle-down effect. This relates to the belief that meaningful changes in rural development would accrue from transferring new technologies to a few selected beneficiaries in turn expected to transfer the innovations downwards and sideways to benefit others. This elitist strategy scored impressive quantitative results at the time of the Green Revolution yet could not prevent widening social gaps as the perculation trickle-down effect did not occur according to expectations. Though differing on account of the scope and context yet not of the analytical relevance, the model villages aimed at by the moshav experiences, failed to generate the kind of spread and reliability that can save a project from its episodic or enclave character. Not only did the costs involved in the settlement of model villages exceed the available resources for a generalizable rural development but the technological package turned out to be too complex and demanding to be effectively decoded by the initial beneficiaries in favor of other groups. This has to be seen as a constraint of the top-down nature of the initiatives.

True, the participatory and frugal behavior of the Israeli experts created a most favorable climate for the emergence of a feeling of shared experience between them and the locals. Yet in their internal communication system, the experts could not refrain from a feeling of superiority of their model vis-à-vis the previous field reality. The project is impressive in its outlook and uniqueness all over Zambia as a cooperative settlement based on the Israeli moshav ovdim. There is an incessant stream of visitors, some of whom are personalities. "Nawalpur is a sort of a show window and ministers and heads of states pay frequent visits to the place . . . many foreign experts arrive and try to learn from our experience . . . their reports testify of our success" (Center for Agricultural Cooperation 1971). Though fully understandable on the part of the then protagonists of action, their postures when seen in retrospective sound different. Today, they invite the criticism directed at the image of the miracle external package and at the vision of linearity of inputs-development-outcome of the external intervention (Long and Van de Ploeg 1989).

Whatever was left of the participatory practices and feelings of comradeship soon waned in the wake of the devolution of the project to the local agencies. It is at that level that the intrinsic top-down ingredient proved the most influential on the future of the projects.

The third insight is the fallacy of the "lower best." The lesson that can be learned here suggests that the presumed viability of the moshav model on account of its relatively lower degree of cooperative integration when compared with the kibbutz and the moshav shitufi proved fallacious. Although lower than the latter, forms still lower than the moshav could be found as early as the 1950s in both Israel and overseas countries. These were the cooperative village and the cooperative within a rural location in Israel (Levi 1975) and the cooperatives of the single/multipurpose types in Third World countries. In other words, the relative advantage of the moshav was inferred through a comparison of the three main forms of cooperative settlements and this a priori restricted its validity.

The main reasons underlying such a relative rather than absolute comparison are obvious. In Israel, it was the success achieved in the pre-State period by all the three forms of cooperative settlements that led to the assumption of the relative advantage of the moshav ovdim in the absorption of new immigrants. This was the stance of the settlement authorities and some sociologists as well, with reservations (Weintraub 1971). As to developing countries, it was mainly the desire to adopt a wholesome organizational pattern through the creation of new impressive settlements. Additivity of dimensions is probably tempting to policymakers aiming at all inclusive and ambitious projects, yet our discussion is rather suggestive of the unpredictability and the multidimensionality of overall settlement viability. In spite of the best intentions of the planners to achieve an all-settlement embracing organization, the settlers' organization, if any, and the settlement community may develop as distinct, or even incompatible entities, as shown by the above Argentinean and Dominican cases.

However, "lower" or easier than the kibbutz and the moshav shitufi, the creation and management of a moshav-type settlement still remains a fairly complex enterprise. This is due to the need to accommodate under the same organizational umbrella a variety of interests and production potentials deriving from the very small holder structure of the settlement. This is compounded by the gap between objectives of the planners and the sociocultural context of the intended beneficiaries. The latter were asked to enter a production and commercial circuit in whose design and implementation they did not participate and which they did not understand. This increased the likelihood of failure.

To conclude, the early infatuation and somewhat naive enthusiasm for the moshav as part of the Israeli assistance overseas and the criticism that came in its wake should be seen in light of the changes that have accompanied the theory and practice of rural development in the Third World over the past three decades. Seen in microcosm, the criticism of the moshav fits into such macro interpretations of contemporary developmental issues as offered by Griffin (1988) and Long and Van de Ploeg (1989). The first analyzed the shift from early infatuation to actual demystification by means of three consecutive stages: the "Brave New World," the "Golden Age," and the "Rude Awakening." Long and Van de Ploeg pleaded for a demythologization of current notions so as to enable us not only to understand but also to transform the practice of planned intervention.

Both interpretations are characteristic of the great shift in current approaches to development issues: from a somewhat naive faith in the potential of Western imported technology to an appraisal of its severe limitations. This, in turn, implies shifting the emphasis from the attempt to help the beneficiaries to conform to the induced innovation to an attempt to learn from their experience and take stock of their wisdom and commonsense toward a possible change. What this chapter aimed at with regard to possible policy implications is that what matters in deciding about an organizational solution for rural beneficiaries in not so much its relative advantage as seen by the planners but rather its relevance to the beneficiaries' broad interests and culture, hence their capability to accept and internalize it as a means of endogenous and pragmatic change. The experience of the moshav overseas may serve as a lesson to policymakers as to the importance of enlisting the participation of the beneficiaries.

## NOTES

1. This view was clearly at odds with the interpretation of Trappe (1966) and Seibel and Marking (1974) who saw in many traditional systems of mutual aid (including the *donkpé*) traditional cooperatives, or even precooperatives.

2. A note on the role of Israeli expertise is in order. In one of the Honduran cases of our study (Levi and Naveh 1988), the Israeli assistance was particularly pragmatic and typically horizontal (i.e., through collaboration between labor movements instead of governments). Although more research is needed to substantiate the advantage of the former over the latter in international development projects, the results seem to contribute a new element to the advantage of participatory strategies over top-down ones.

## REFERENCES

Braverman, A., J. L. Guasch, M. Huppi, and L. Pohlmeier. 1991. Promoting Rural Cooperatives in Developing Countries: The Case of Sub-Saharan Africa World Bank Discussion Paper 924 No. 121. Washington, D.C.: The World Bank.

Center for Agricultural Cooperation. 1971. Bulletin No. 5 (March). Rehovot: Ministry of Agriculture and of Foreign Affairs, Center of Cooperation and Agricultural Development in Developing Countries.

Desroche, H. (with the collaboration of Zwi Gat). 1973. Opération Moshav: D'un Développement des Villages à une Villagisation du Développement. Paris: Ed. Cujas.

Eicher, C. K. 1969. "Reflections on Capital-Intensive Moshav Farm Settlements in Nigeria." In *Agricultural Cooperatives and Markets in Developing Countries*, K. R. Aschel et al., eds. New York: Praeger.

Frank, M. 1968. *Cooperative Land Settlements in Israel and their Relevance to African Countries.* Basel: Kyklos Verlag.

Gil, M., and F. Oliart. 1980. Las Empresas Campesinas de Reforma Agraria (ECRA) en el IAD. Santo Domingo (photocopy).

Gosselin, G. 1970. *Développement et Tradition dans les Sociétés Rurales Africaines.* Genève: Bureau International du Travail.

Griffin, K. 1988. "Thinking about Development: The Larger View." *Development Journal of the Society for International Development* (2–3): 5–8.

Hubkapong Cooperative Village (under the Royal Patronage). 1982. Bangkok: Cooperatives' Promotion Department, Technical Division (photocopy).

International Labor Organization. 1967. Rapport au governement de la République du Dahomey sur la Structure du Service de la Coopération, la Formation et l'Education Coopérative et le Statut de la Coopération. Genève: OIT, Document polycopié.

Kahan, D. 1986. "Institutional Development: Some Lessons from the Experience of Israeli Technical Assistance in the Setting Up of Moshav Multi-Purpose Cooperatives." *Agricultural Administration* 21: 135–146.

Kaunda, K. 1967. *Zambian Humanism.* Lusaka.

Kreinin, M. 1962. "The Introduction of Israel's Land Settlement Plan to Nigeria." *Journal of Farm Economics* 45: 535–546.

———. 1964. *Israel and Africa: A Study in Technical Cooperation.* New York: Praeger Special Studies.

Levi, Y. 1970. "Assistance in the Promotion of Cooperative Activities." In *Israel–Africa Cooperation Research Project* (Progress Report), Y. Herschlag, ed. Tel Aviv: Tel Aviv University, Department of Developing Countries.

———. 1975. "A Methodological Framework for the Classification of Rural Cooperatives in Israel." *Journal of Rural Cooperation* 3(2): 119–129.

———. In press. "Planning for Beneficiary Organizations in Resettlement Projects and the Fallacy of the Lower Best." *International Social Science Review.*

Levi, Y., and Naveh, C. 1988. *Towards Self-Management in New Land Settlement Projects: A Cross National Study.* Boulder, Colo.: Westview Press.

Long, N., and J. D. Van de Ploeg. 1989. "Demythologizing Planned Intervention: An Actor's Perspective." *Sociologia Ruralis* 29(3–4): 227–249.

Meister, A. 1966. L'Afrique Peut-Elle Partir? Paris: Seuil. Ministry of Agriculture; Rehovot: Center for Agricultural Cooperation with Developing Countries (file Cameroon, photocopy).

Nyerere, J. K. 1962. *Ujamaa: The Basis of African Socialism.* Dar-Es-Salaam: Tanu.

Oluwasanmi, M. O. 1965. L'Adaptation de Structures Coopératives Modernes au Développement de l'Agriculture Nigérienne. Archives Internationales de Sociologie de la Coopération, No. 18.

Oni, S. A., and J. K. Olayemi. 1973. "Determinants of Settler Success under the Western Nigerian Farm Settlement Scheme." *Nigerian Agricultural Journal* 10: 228–239.

Onouha, B. 1965. *The Elements of African Socialism.* London: André Deutsch.

Prives, M. Z. 1977. The Development Projects in the Terai of Nepal. Rehovot: Center for Agricultural Cooperation with Developing Countries (photocopy).

Ralston, L., I. Anderson, and E. Colson. 1983. *Voluntary Efforts in Decentralized Management: Opportunities and Constraints in Rural Development.* Berkeley: University of California.

Reiss, A. 1969. Pilot Settlement Projects in Nepal. Rehovot: Center for Agricultural Cooperation with Developing Countries (photocopy).

Sawadogo, R. C. 1979. *Les Migrations: Problemes des Etablissements Actuels et Elements d'approche pour une Definition des Politiques de Colonisation des Terres Neuves dans la Region du Programme de lutte Contre l'onchocercose.* Ouagadougo: World Health Organization.

Scudder, T. 1981. The Development Potential of New Land Settlement in the Tropics and Subtropics: A Global State-of-the-Art Evaluation with Special Emphasis on Policy Implications. Pasadena: Institute for Development Anthropology and California Institute of Technology.

Seibel, H., and H. D. Marking. 1974. *Traditional Organizations and Economic Development Studies in Indigenous Cooperatives in Liberia.* New York: Praeger.

Senghor, L. 1961. *Nation et Voie Africaine du Socialisme.* Paris: Présence Africaine.

Texier, A. F. 1969. "Institutional Problems in the Modernization of African Agriculture." In *A Review of Rural Cooperation in Developing Areas.* Geneva: United Nations Research Institute for Social Development.

Trappe, P. 1966. *Entwicklungsfuncktion des Genossenschaftswesens am Beispiel Ostafrikanisher Stamme.* Berlin: H. Luchterhand.

Tsiranana, P. 1962. *Programme Economique.* Tananarive: Imprimerie Nationale.

Uphoff, N. 1987. Relations between Governmental and Nongovernmental Organizations and the Promotion of Autonomous Development in Promotion of Autonomous Development. Report of the Proceedings of the Expert's Consultation held at Noordwijk, The Netherlands, October 27–30.

Van Raay, M. G. T., and J. G. M. Hilhorst. 1981. Land Settlement and Regional Development in the Tropics: Results, Prospects and Options. Draft Discussion Paper. The Hague: The Institute for Social Studies Advisory Service.

Verhagen, K. 1984. *Cooperation for Survival.* London: International Cooperative Alliance and Amsterdam: Royal Tropical Institute.

Weintraub, D. 1971. *Immigration and Social Change, Agricultural Settlement of New Immigrants in Israel.* Manchester: Manchester University Press.

Weitz, R., D. Pelley, and L. Applebaum. 1978. Employment and Income Generation in New Settlement Projects. World Employment Paper 10, Working Paper 3. Geneva: International Labor Office.

Weitz, R., and A. Rokach. 1968. *Agricultural Development Planning and Implementation (Israel Case Study).* Dordrecht, Holland: Reidel Publishing Co.

World Bank. 1985. Sustainability of Projects: First Review of Experience. Operator's Evaluation Dep., Report No. 5718. Washington, D.C.: World Bank.

Zafrir, I. 1980. Informe de las Actividades Realizadas (Junio 1978–Junio 1980) Santo Domingo, 1980. Issued by Republica Domicana, I.A.D. Division de Asistencia Tecnica y Estado de Israel, Min. de Relaciones Exteriores, División de Cooperación Internacional, 1980.

Chapter 13

# The Decooperativization of Israel's Moshavim, 1985–1994

*Moshe Schwartz*

*Throughout much of the short history of the moshav as a major rural institution in Israel, it has occupied a special relationship with the state. While being used as an instrument of social change (such as a device for the absorption of immigrants), it has also enjoyed economic protection from the vicissitudes of the general Israeli economy. In this chapter, Moshe Schwartz considers the impact of a reduction in this protection upon what some consider to be fundamental to the character of the moshav: its commitment to a cooperative economic structure.*

Following an unprecedented financial crisis, far-reaching changes have occurred in the past ten years (1985–1994) in the economic, political, and organizational environments of Israel's 400 or so small holder cooperative settlements, the moshavim. The government no longer provides unlimited (albeit unwritten) financial guarantees to the settlement sector. Almost all of the national and regional moshav organizations have become insolvent and powerless; and in the years 1987 to 1991, agriculture lost much of its importance in the national economy.

As a result of these changes in their environments, a rapid transformation has taken place in the moshavim. This chapter reviews major aspects of the transformation—decooperativization, changing occupational structure, and in parts of the country, suburbanization and gentrification. To give a comparative perspective, we also refer to some changes over the same period in Israel's collective settlements, the kibbutzim. We suggest a possible explanation for the transformation in terms of the partial removal of the barriers between the moshavim (as well as the kibbutzim) and the general economy of the country in each of the four markets of classical economics: commodities, labor, capital, and land.

While change in the rural sector has been slowing down since 1992, there are reasons to believe that moshavim may once again undergo a period of rapid change, this time with respect to the property rights of their membership.

## FINANCIAL CRISIS AND THE FLIGHT FROM MUTUAL GUARANTEES

In the summer of 1985, three months before a financial crisis of unprecedented proportions erupted in the settlement organizations, escalating inflation, reaching a peak of 500 percent per annum, was threatening to inflict irreversible damage on Israel's economy. The new National Unity Government saw the curbing of inflation as its foremost task, and, together with the Central Bank, it took vigorous measures, including a wage-and-price freeze, steep real interest rates, and drastic restriction of credit. In three months, inflation was down to 60 percent per annum. This success entailed heavy costs. Indebted businesses saw their debts soar while their ability to keep functioning was severely curtailed. Not a few of them went under. Because the agricultural settlement sector, which includes the moshavim and the kibbutzim, was heavily in debt, the government's new policies could not but affect it strongly.

Much of the debt of the moshavim was to their purchasing cooperatives, organizations originally set up to supply the moshavim with cheaper agricultural inputs such as animal feed and fertilizer. With the passage of time, many of these cooperatives had expanded their activities to include processing of those inputs and of the produce of the moshavim and providing them with lobbying services and credit. Most purchasing cooperatives had become powerful combinations of agribusiness concerns, political machines, and banking operations. The moshavim and the purchasing cooperatives were linked to each other by unlimited financial guarantees. By 1985, the purchasing cooperatives were heavily indebted to the country's banks. However, they were not too worried about their situation. Up to now, whenever such debts had threatened the ability of the purchasing cooperatives to function, they had been bailed out by the government and by the Settlement Department of the Jewish Agency, a quasi-governmental agency that had been in charge since pre-State days of planning and establishing new rural settlements, endowing them with appropriate means of production and assisting them until it pronounced them able to function independently (for additional information, see Chapters 2 and 3).

In September 1985, one moshav purchasing cooperative, unable to honor its financial obligations, appealed as usual to the government for help. For the first time, the government refused assistance, and its indirect financial aid, squeezed out of paragovermnental organizations such as Agrexco (the national agricultural export company) and the Flower Council by Ministry of Agriculture officials (Government Comptroller 1988) was too small to make a difference. The purchasing cooperative had to go into receivership.

Heavy debts amounting to about $600 million were by that time burdening most purchasing cooperatives. More than half of that sum was owed to Israel's two largest commercial banks, both of which had counted on the government to prevent cooperative bankruptcy. The government's refusal to bail out the cooperatives caught the banks by surprise. From then on, they rejected the requests of the purchasing cooperatives and their member moshavim to recycle the loans.

Because annual inflation was still above 50 percent, the Bank of Israel and the treasury kept the credit market tight and real interest rates very high (40% to 60% on the average). The debts of the purchasing cooperatives swelled. The Finance Committee of Israel's Knesset voted to grant them $150 million emergency assistance, and the Jewish Agency handed them $50 million, all to no avail.

The purchasing cooperatives, most of them regional, were secondary cooperatives, which means that their clients and members (and thus also owners) were moshavim, not individual farmers. There were unlimited financial guarantees operating between every supply cooperative and its member moshavim, as well as between every moshav and its member small holders. The purchasing cooperatives had financial ties to one another and were involved together in large commercial operations (such as importing grain for the manufacture of animal feed). Most purchasing cooperatives also had ties of mutual guarantees with the economic organizations of the Moshav Movement (such as its insurance company, which had succumbed in 1985 to its enormous debts).

The system of mutual financial guarantees operated as follows: When a purchasing cooperative wanted to get a loan from the bank, often to redistribute it among its member moshavim, it would bring as collateral a sheet of paper bearing the signatures of all the member moshavim as guarantors of the loan—those sheets were labeled "bed sheets" because of their length). With time, all moshavim in any given purchasing cooperative became guarantors of all its debts. Thus, if one moshav failed to pay its debt to the purchasing cooperative, the others had to pay it. In effect, the guarantees of the member moshavim to the purchasing cooperative operated as a system of mutual guarantees between those member moshavim. The system was also operative at the level of the individual moshav (i.e., when the moshav wanted a loan from the bank or from the purchasing cooperative, it would bring as collateral a sheet of paper bearing the signatures of all individual members). In effect, the guarantees of individual members to their moshav association operated as a system of mutual guarantees between members of each moshav.

With the help of this system, the moshavim had been able to engage in credit intermediation on a very large scale, obtaining loans for their members and also lending the money that some members had deposited with them against high interest rates. The same type of arrangement worked between the moshavim and the purchasing cooperatives, with some moshavim lending money to the purchasing cooperative and most of them borrowing money from them. Thus, the mutual guarantees system allowed an enormous amount of borrowing by moshav members, moshavim, and purchasing cooperatives. The banks "knew," as mentioned earlier, that if the purchasing cooperative defaulted on its debts, the state would eventually pay for it. In other words, unlimited, if unwritten, financial guaranties by the state stood behind the willingness of the banks to accept mutual guarantees as collateral.

By September 1986, ten out of twelve regional purchasing cooperatives had become insolvent. Their aggregate debt (including that of member moshavim) had risen to $1.5 billion. Many moshavim were unable to meet interest payments on their debts, let alone repay the principal. Others worried about their indirect debt, incurred through guarantees to defaulters. Indeed, given the system of mutual guarantees, the debt threatened to bury the moshav sector.

The Israeli government was obliged to demonstrate concern for the beleaguered moshavim. The settlements, after all, still retained some remnants of their pioneer prestige, and their organizations still wielded some power. Besides, they continued to produce $1 billion worth of agricultural goods annually, and the government, now owner of the banks, was worried about their financial standing.

Several commissions were appointed by the government to find out how the moshav purchasing cooperatives had become insolvent, to rescue the "family farm sector," and to devise safeguards against breakdown recurrence. In 1987, the government, the Jewish Agency, and the banks drafted an agreement designed to enable the moshavim to pay their debts and still remain economically viable. Debt payments were to be rescheduled and extended over a number of years so as not to exceed the payment capacity of every moshav and moshav member. It was also agreed that those unable to pay would receive a partial moratorium, provided that they give up their small holdings and production quotas. However, few of the moshavim signed the agreements, and the debts of the others kept on mounting.

Other major debtors of the country's two largest banks were the kibbutzim (including most kibbutz purchasing cooperatives) and other enterprises affiliated with the Histadrut, Israel's powerful labor union organization. Most of these debtors were on the brink of bankruptcy. Furthermore, the banks themselves were heavily in debt because of a severe crisis in October 1983, as a result of which the government had bought 75 percent of their shares.

As a matter of fact, preventing moshavim, kibbutzim, and Histadrut enterprises from going under was one of the motives that in the late 1980s induced the Labor Party to join a Likud-headed government so that then leader of Labor, Shimon Peres, might head the treasury and help labor-affiliated institutions—moshavim, kibbutzim, and Histadrut enterprises—out of trouble.

None of this assistance, however, could breathe new life into the collapsed purchasing cooperatives. Indeed, if labor had tried to revive them, even with a revised system of carefully restricted mutual guarantees, the moshav membership would have rejected the attempt. So profound was the trauma that the moshav small holders now refused to provide their fellow moshav members with any financial guarantees even though some banks were willing to offer much cheaper credit to moshavim where the membership still provided financial guarantees to each other.

## CHANGES IN THE MOSHAV'S ORGANIZATIONAL ENVIRONMENTS

### Decline of Other Support Organizations

The purchasing cooperatives of the moshavim were not the only moshav organizations whose existence was terminated or at least seriously threatened by the crisis. Another endangered organization was the Moshav Movement, Israel's largest moshav federation and lobby. The insolvency and incapacitation of the purchasing cooperatives deprived it of an essential source of operating funds. Since the Movement's own economic organizations went under even before the purchasing cooperatives, it had to reduce its manpower and to rely for funds on the Histadrut, by now also in dire straits. Besides, the Movement had already lost much of its legitimacy (Sherman and Banin 1992), and member moshavim would not dream of paying dues.

The Settlement Department of the Jewish Agency no longer exists as an independent body. Its manpower and budget have been severely reduced, and what is left of it has been united with the remnants of the Renewal Department of the Jewish Agency to form the new Department of Development and Settlement, which has no financial commitment to the individual or collective survival of any individual settlement or settlement form and concentrates its effort in the promotion of regional development in peripheral rural areas within the Green Line.

The Ministry of Agriculture is now trying to take over some of the earlier functions of the Settlement Department and has renamed itself the Ministry of Agriculture and Village Development. However, its budgets and personnel have shrunk more during the last fifteen years than those of any other government ministry (Sharkansky 1988).

## Decline in State Support for Agriculture

The moshavim (as well as the kibbutzim) were set up by national capital and helped during their first years. Even later on, whenever their continued existence was threatened, the state and the Jewish Agency would step in and provide them with additional financial assistance. The ongoing regular aid and the recurrence of emergency assistance convinced lenders (banks, input sellers, and moneylenders) that the settlements enjoyed state insurance against bankruptcy and that credit could be extended to them without too much concern about their ability to repay their debts.

In the 1950s and early 1960s, credit was strictly rationed in Israel so that the state guarantees against bankruptcy only caused limited damage because the settlements could only run a limited amount of debts. For example, the kibbutz sector, which had higher debts than the moshav sector, was extricated in the early 1960s from its troubles with the relatively moderate sum of $30 million (even taking into account dollar depreciation, this is far less than the more than $3 billion or more owed by the kibbutzim thirty years later).

After the Six Day War and in the 1970s, credit became much looser and Israel's big banks started to look for customers to take their money. The unlimited financial guarantees enjoyed by the rural sector came in handy. From now on, it started to obtain unlimited credit from the commercial banks on the basis of the unwritten insurance it received from the state and the Jewish Agency that they would not let it down. As a result of this new possibility, debts started to pile up much faster than in earlier years. The ratio between debt and productive assets grew by a factor of 4.2 between 1969 and 1986 as compared to a factor of 1.22 for industry (Zussman, Kislev, Lerman, Kroll, and Rosenthal 1990). This process eventually culminated in the unprecedented financial crisis of 1985/86, with the subsequent loss of the financial life insurance of the state and of the special standing of the settlement sector in the financial markets.

The volume of credit to the agricultural sector has been declining in real terms. For example, during 1983, credit to agriculture grew only by 14 percent in nominal terms, while inflation was higher than 15 percent. Furthermore, the decline of the share of agriculture in the overall amount of credit received by the public is a long term trend: In 1988, agriculture received close to 15 percent of the total amount of credit extended to the public, whereas in 1994, the share had gone down to less than 7.5 percent (Ministry of Agriculture 1994). It should also be mentioned that in 1988 already much of the credit to agriculture was involuntary credit, given because organizations and settlements failed to repay their debts and the bank had not yet written them off.

Moshav members are now treated with suspicion when they ask for loans, and in a curious reversal of the precrisis situation, they now may well have to provide the banks with more securities than other borrowers (personal interview with an official at one of Israel's large banks, 1994).

### Decline of Settlements in the Public Opinion of the Country

The settler, kibbutz or moshav member, was one of the culture heroes of the Yshuv (Jewish pre-State society in Palestine) and played an essential role in the establishment of the state of Israel. Settlements defined the borders of the newly established state, and settlement members were the backbone of its military. In the 1950s and early 1960s, the agricultural sector continued to be perceived as a success story. Thus, it managed to bring the country from extreme food scarcity in 1948 to self-sufficiency and even to agricultural surpluses in 1958 even though the population had tripled (Schwartz and Giladi 1993). The agricultural sector, and especially the moshavim, also set an impressive record in the domain of immigrant absorption. High productivity levels and technological advances strengthened the positive image of Israeli agriculture.

Even later on, in the 1960s, 1970s, and even in the 1980s, people still looked up to the members of agricultural settlements as contributing disproportionately to the country's security as well as to its economy, whether through industrial achievements, for which the kibbutz was famous, or through agricultural exports, in which the moshav played an eminent role.

Once the financial crisis hit the settlement sector, the settlers were suddenly looked upon as a group that did not pay its debts and lived off taxpayers' money, which also had to be used to bail the settlers out. The assistance that labor had helped make available to the agricultural settlement sector had eventually become a political liability.

### Decline of Agriculture

The share of agriculture in the country's economy and in the employment of its labor force underwent slow decline before 1986 and more rapid decline thereafter (see Table 13.1). The share of agriculture in the country's gross national product (GNP) declined from 12 percent in 1952 to 8 percent in 1966, 5.1 percent in 1987,

**Table 13.1**
**Share of Agriculture in the GNP**

| Year | Share in GNP |
|---|---|
| 1952 | 12% |
| 1966 | 8% |
| 1987 | 5.1% |
| 1989 | 4.1% |
| CRISIS | |
| 1990 | 3% |
| 1991 | 2.5% |

4.1 percent in 1989, 3 percent in 1990, and 2.5 percent in 1991. In the five years between 1986 and 1991, the share of agriculture in the country's GNP declined by 50 percent, more than in the previous twenty years.

Employment in agriculture decreased from 17.6 percent of the labor force in 1955 to 12 percent in 1966, 8.8 percent in 1970, 6.4 percent between 1975 and 1980, 5.2 percent in 1986 and 1987, 4.6 percent in 1988, 4.2 percent in 1990, and 3.5 percent in 1991. Around the same time (between 1987 and 1991), the share of agriculture in the labor force declined by 33 percent, almost as much as in the previous seventeen years, 1970 to 1986 (see Table 13.2).

The number of those engaged to a significant extent in agriculture fell from 121,100 in 1959/60 to 89,800 in 1960/70, 87,700 in 1979/80, and 87,700 in 1986/87. After the onset of the 1985 crisis in cooperative organizations and settlements, the numbers dropped to 79,600 in 1988, 78,000 in 1989, 70,900 in 1990, and 64,300 in 1991. Thus, in the five years from 1986 to 1991, the number of agriculturists decreased by 27 percent, almost the same as in the previous twenty-five years (28%). It should be stressed that,

**Table 13.2**
**Share of Agriculturists in the Labor Force**

| Year | Share in labor force |
|---|---|
| 1955 | 17.6% |
| 1966 | 12% |
| 1970 | 8.8% |
| 1975–80 | 6.4% |
| CRISIS | |
| 1986–87 | 5.2% |
| 1988 | 4.6% |
| 1990 | 4.2% |
| 1991 | 3.5% |

Data from Central Bureau of Statistics 1993.

as in most Western countries, it was not the quantities of agricultural produce that decreased but rather their price per unit on the market. As a matter of fact, productivity went up as a shrinking population of agriculturists produced more and sold it for less. Thus, agricultural production rose by about 6 percent between 1992 and 1993 but fetched 4 percent less in real terms.

The decline of the share of agriculture in the national economy has not been accompanied by rural depopulation. On the contrary, over the past thirty-odd years, the countryside has generally maintained its share in the rapidly expanding population of the country. The only exception was during the 1960s, when the share of the countryside in the overall population dropped from 13 percent in 1961 to 9.8 percent in 1972, and even in those years the rural population grew in absolute numbers from 251,800 in 1961 to 263,800 in 1972. Thereafter, the share of the rural population in the total population remained stable, and in the years 1985 to 1989, it even rose above the 10 percent mark. It is only because of the boost in urban population due to immigration that the rural population dropped to 9.7 percent in 1991.

The steady growth of the rural population, as manifested by its unchanging share in the total population, was made possible by the continuous expansion of the nonfarm population in the moshavim

**Table 13.3**
**Number of Individuals Engaged in Agriculture**

| Year | No. engaged in agriculture |
|---|---|
| 1959–60 | 121,100 |
| 1960–71 | 89,800 |
| 1979–80 | 87,700 |
| CRISIS | |
| 1986 | 87,700 |
| 1988 | 78,000 |
| 1990 | 70,900 |
| 1991 | 64,300 |

(as well as in the kibbutzim) and more broadly by the processes of urbanization and suburbanization, which are one of the topics of the next chapter.

## TRANSFORMATION OF THE MOSHAVIM

### Decooperativization

The moshavim have been deeply affected by the changes in their environment, and one of their responses is reflected in the reduction of their cooperative features. The process is not new, as it has been going on at least from the time of the establishment of the state. In the moshav, decooperativization has expressed itself in the flight from the communal marketing of agricultural produce, from mutual assistance, and, for those members who found it possible, from exclusive dependence on the moshav association for credit purposes. In an important agricultural branch that grew from the late 1970s on, the flower branch, the relevant regional and national authorities regulating the branch, dealt directly with the members without the association as intermediary.

The acceleration of decooperativization was part of a chain reaction following the strict antiinflationary policies adopted by the Bank of Israel and the treasury in the summer of 1985. Subsequent events, particularly the collapse of the purchasing cooperatives of the moshavim, fatally undermined the confidence of the moshav membership in the system of mutual guarantees. Then, the first moshav purchasing cooperative of moshavim fell, dramatizing the abolition of the unwritten and unlimited guarantees between moshavim and individual moshav members. Whereas formerly state backing was taken for granted, from now on there was an actual danger that as a guarantor the member might have to pay all the debts of other members of his or her own and other moshavim. These circumstances contributed to increasing unwillingness among settlement members to act as guarantors for the debts of other members and to guarantee the debts of other settlements. This was the start of the rapid decooperativization of the moshavim.

Only a small number of moshavim are still being run according to the rules in effect a decade ago and continuing to maintain mutual guarantees and communal marketing. In all the rest, even in veteran moshavim, once characterized by high levels of mutual trust, the system of mutual guarantees has been abolished, giving way to an attitude of everybody for themselves (Schwartz, 1995).

On many moshavim, all economic cooperation between the members has ceased. Most of them can no longer operate as economic units so that these settlements now function as ordinary villages. A large group of moshavim has even lost the ability to deliver minimal municipal services to its members, and an increasing share of that task has been taken over by the regional councils.

### Changing Income and Occupational Structure

Additional changes in the moshavim are the decline of agriculture as a source of income, the growth of alternative sources of revenue (such as rural tourism [Fleischer, Banin, and Kahana 1991] and other small business operations [Sherman and Kedar 1993; Sherman, Laks, and Kedar 1994]), and an increase in the proportion of moshav members who work outside the moshav (outside workers who often do not derive any part of their income from agriculture).

As early as the 1950s, when agricultural employment reached its peak among the Jewish population of Israel, there were some moshav members for whom agriculture was not the only source of income. This was even the case in veteran moshavim, those set up before the establishment of the State of Israel, where commitment to agriculture was very strong and where the size and capital intensiveness of the farm were on the average much higher than in moshavim established later. In the 1960s, the rate of outside workers among moshav members increased. The rate of moshav members employed exclusively in agriculture dropped from 75 percent in 1962 to less than 60 percent in 1972, and until 1980, it further declined to 51.5 percent. The fruit tree branch, especially appropriate for part-time farmers, became the largest in the moshav economy, and between 1971 and 1981, the numbers of fruit growers doubled (Shklanewicz 1983).

### Urbanization

As already mentioned, the growth of the population of the countryside and the maintenance of its share in the total population has been made possible by the suburbanization of the moshavim situated close to the large cities of the country. In the 1970s, the moshavim located near those cities became functional equivalents of suburbia with something extra. On the one hand, they were increasingly attractive to young city dwellers seeking to improve their "quality of life" (Applebaum 1986) and unable to afford a one-story house with a lawn in one of the new middle-class satellite cities around Tel Aviv, Haifa, or Jerusalem. On the other hand, they even offered something extra: life in a much smaller family-like community than the satellite cities, one in which dwellers could have more influence over their environment, and also one in which they could practice some part-time farming if they so wished. The population of the moshavim grew from 125,000 in 1972 to 143,000 in 1980 and to 152,000 in 1991. This growth was not due to an increase in the number of households, since at that time, a new family could enter the moshav only by purchasing an existing farm, whose previous owners would then have to leave the moshav. However, the new families were often young and the women of childbearing age. In those years, the proportion of part-time farmers grew considerably (Shklanewicz 1983).

In the 1980s, additional changes brought the population of the moshavim even closer to that of nearby cities. Very rapid inflation and, later on, the financial crisis of the agricultural settlement sector reduced the population of farmers, and since that of the countryside kept increasing and few exfarmers left, it is quite clear that the proportion of part-time farmers and nonfarmers in the rural population continued to grow.

A further change in the offing is a moshav expansion program, the *harhavot,* decided upon in 1990 by the planning institutions of the state of Israel as well as by the Lands Authority with an eye to providing some financial assistance to the embattled moshav membership, increasing the population of the countryside, and coping with the severe housing shortage aggravated by mass immigration from Ethiopia and the Soviet Union. Prior to the decision on the harhavot, only the moshav member himself or herself and the designated heir of the small holding ("continuing son or daughter" in the official moshav terminology) could build a house on the member's plot of land and be entitled to reside in the moshav. According to the new plan, this restriction has been partly lifted and an additional "noncontinuing" sibling can build a house, thus allowing for a doubling of the population of every moshav whose members decide to embrace the plan and permitting each moshav family to help another of their children. Some of the harhavot houses can also be purchased by outside persons. The new plan will further dilute the agricultural population of the moshav. It will also allow more Israelis to realize the national middle-class dream of quality of life (a single-family house with a lawn and garden).

Additional consequences of the harhavot plan for the moshavim might be increased variety and rejuvenation of the moshav population, as well as acceleration of the gentrification already occurring in Israel's countryside, as in that of most other Western countries. The new policy might also provide economies of scale in the delivery of municipal services and make for cheaper or better services.

By August 1991, more than half of the 400 or so moshavim in the country had decided to embark on the harhava process and more than 20 percent had already submitted their plans to the local or district commissions, but only a few had reached the implementation stage (Applebaum and Kedar 1992). One year later, 60 percent had applied to join the process, and 40 percent had submitted plans, most of them in the coastal zone or in the center of the country (Applebaum and Kedar 1993). To date, new houses have been built in only a handful of moshavim.

An interesting question is why changes in the moshav (as well as parallel changes in the kibbutz) that crept along slowly for more than twenty-five years underwent such a marked acceleration after 1986. A related question concerns their slowdown after 1991.

One possible explanation is that barriers that partly separated the settlement sector from its economic environment have been lowered in the wake of the financial crisis, increasingly weakening the historical linkage between village, agriculture, and collective-cooperative arrangements and between place of residence and place of work. This weakening increased the similarities of the rules of the game in the settlement sector with those of the rest of the Israeli society (Schwartz and Applebaum 1994).

### Institutional Barriers between the Settlement Sector and Its Environment

In the first decades of statehood, cooperative and collective agriculture was to a large extent separated from the rest of Israel's economy and society by insulation mechanisms such as laws and organizational systems. Some results of this separation have been documented (Vilkansky 1980), and the phenomenon went so far as to induce a lack of relationship between the social and economic situation of a village and that of the nearby town. There were flourishing villages—kibbutzim and moshavim near development towns that did not develop, such as Kiryat Shmona and Beit Shean, and there were, on the other hand, weak villages near well-developed towns, such as Netanya. The insulation mechanisms made for a sharp differentiation between city and village, despite the small size of the country. They also helped maintain the original organization of the kibbutz and the moshav and served as shock absorbers against the sharp and rapid changes in the Israeli economy and society. In addition, they slowed down development by protecting the settlements from the need to adapt to their changing environment.

The economic separation between the settlement sector and the rest of the country can be analyzed in terms of the four markets of classical economics—commodities, capital, labor, and land.

## The Separation between the Sectors in the Commodities Markets

In the commodities markets, there were two domains of separation, one pertaining to the settlements as sellers and the other one as buyers. The kibbutzim and the moshavim sold their produce via bodies they owned, such as *Tnuva* (the national cooperative that markets much of the country's agricultural produce, including most of the milk produce and a large proportion of the poultry produce) and the regional plants that process agricultural produce. At the same time, kibbutzim and moshavim purchased agricultural inputs and other products through other organizations they owned, such as the regional purchasing cooperatives.

The economic separation of the settlements from the rest of the country in the commodities markets was never complete. Kibbutzim and moshavim also purchased commodities on the private market when they stood to gain by doing so. They also sold commodities, especially vegetables and fruit, through private merchants when the latter offered higher prices or better payment conditions than Tnuva.

The fact that the settlement sector had access to sale and purchasing organizations of its own provided it with economies of scale but also caused it to bear the functioning costs and, eventually, the deficits of those organizations. Under the new conditions that emerged after the financial crisis, the settlements and subsectors within the settlement sector lost much of their ability to cooperate and share the costs with each other. Thus, they are hardly able to maintain commercial organizations of their own, regional or national. Some of the organizations that operated in the mid-1980s have gone under or have been taken over by private capitalists.

## The Separation between the Sectors in the Capital Market

In the capital market, the separation between sectors is the historical result of the fact that collective and cooperative settlements were set up and generously assisted by national capital, as already described. This lasted until the unprecedented financial crisis of 1985/86, with subsequent loss of the financial insurance of life by the state (I have used the term *insurance of life* rather than *life insurance*. Life insurance insures your inheritors of some compensation in case you die. The state insured instead eternal financial existence to the settlements). The settlement sector then lost its standing in the financial markets. It became harder for kibbutzim and moshavim to borrow money at the bank than it was for ordinary

Israeli citizens. From 1987 onward, there also was a marked reduction in government support of agriculture, both direct and indirect (Schwartz 1995).

## Separation between the Sectors in the Labor Market

In the labor market, an essential factor in the separation of the settlement sector from the rest of the country was the requirement on the part of the settlement institutions that the kibbutzim and the moshavim live off the means of production allocated to them by those institutions. Moreover, kibbutzim and moshav members were required to use these means of production by their own labor and avoid employing salaried workers. In the kibbutzim, the norm was that the members work in one of the various kibbutz branches unless the kibbutz itself sent them as its envoy to some external organization. The moshavim (especially the older ones) demanded that members derive their livelihood exclusively from agriculture, or at least make efforts in that direction, although the requirement was never enforced (Schwartz 1986). In fact, however, the principle of self-labor was not generally put into practice in the moshavim, and even in the kibbutzim, it was sometimes honored only in the breach.

The overall result in terms of the separation between the sectors in the labor market was twofold. On the one hand, settlement members were assured of an advantage on the internal labor market; they enjoyed a preferential status in the settlements and the regional plants similar to the preference enjoyed by the owner's family in family enterprises in matters such as being hired, advancement, and immunity from firing. On the other hand, administrative regulations made it harder for settlement members than for the rest of the population to enter the general labor market through official channels. Settlement members were denied free access to the regional labor exchanges of the Ministry of Labor and Welfare, according to an agreement entered into by the settlement movements and the Ministry in the early days of statehood (Schwartz, Bar-El, Nesher, and Finkel 1987).

The separation of the settlement sector from the rest of the country with regard to the labor market could not possibly remain in effect in the long-run. Full employment allowed almost anyone to get a job without the help of labor exchanges, and indeed, the rate of outside labor among the membership of the moshavim increased over the years and so did the rate of hired labor in the moshavim. Factors contributing to the latter situation were the availability of cheap labor from the territories occupied in the Six Day War and the expansion of labor intensive branches such as viticulture and greenhouse farming of vegetables and flowers.

A breach in the administrative separations between the sectors in the labor market opened in the early 1980s while Aaron Uzan, himself a moshav member, was Minister of Labor. Uzan decreed that the moshav membership be allowed free access to the labor exchanges. In fact, the implications of this innovation were mainly symbolic. A number of factors made it difficult for moshav members to become integrated in the general labor market of Israel. These included the expectations held by many of them that the state and the Jewish Agency would go on providing them with a living and their insufficient professional training and the fact that they saw themselves as small independents rather than as salaried workers. Besides, moshav members had free access to the credit extended to their settlement by the regional purchasing cooperative and passed on to them so that they could buy on credit not only agricultural inputs but also consumption goods at the local village store. This situation ensured their basic existence and may have removed some of the incentive for seeking outside jobs.

However, even after the collapse of the purchasing cooperatives stopped the free access to credit, the moshav members still did not hurry to the labor exchanges to look for jobs (Schwartz 1986). Some probably found ways to make a living without entering the official labor market; others have probably been living off their meager capital.

To some extent, the settlement sector still constitutes a separate labor market. Almost 90 percent of kibbutz members still work in their kibbutz (as mentioned, only 11 percent work outside), and almost half of the moshav membership still does some farming. On the other hand, outside work is on the increase in both kibbutzim and moshavim, while the internal labor market is not growing with time. Furthermore, the preferential treatment enjoyed by the kibbutz and the moshav membership in that internal labor market is weakening. In some kibbutzim, a branch coordinator or plant manager already can fire a member, and it frequently happens that a kibbutz prefers to appoint a nonmember to a managerial post so as to get the best person for the job and also to prevent special consideration toward fellow kibbutz members based on communal ties.

### Separation between the Sectors in the Land Market

The separation in the land market has shown greater endurance and is more stable than the other separations between the settlement and the urban sectors. It is ensured by the legal distinction between agricultural and other land and also by the use of legally binding written contracts such as land leases and sales. The separation is also rooted in the ownership structure of the land market: Indeed, for all practical purposes, 90 percent of the land in Israel is owned by the state or by the Jewish National Fund, a ninety-year old organization formerly in charge of purchasing land for the Zionist movement and still the owner of very large tracts of land. This combination of legal distinctions and ownership structure has made the separation between the sectors stronger and more stable in the land market than in the other three markets where the separations were based on the more tenuous infrastructure of ideological requirements, administrative regulations, and shifting organizational interests.

The legal features of the separation in the land market were planned in such a way as to protect the agricultural village and maintain Jewish control of the national land. However, the protection of agricultural land has weakened in recent years. This weakening has found its expression in the authorization to build new housing units (harhavot) in the moshavim and in the waning of the veto power of the Commission for the Preservation of Agricultural Land as a factor in the decisions made by the commissions of planning and building. In addition, a new situation has been created following the adoption of a resolution by Israel Lands Authority (Resolution 533 of 1992, replaced by Resolution 611 in 1993 [Council of the Lands of Israel 1992, 1993]) that helped to facilitate the transformation of agricultural land into land on which, under certain conditions, people are authorized to build dwellings or industrial buildings. Through these changes, the government was attempting to respond to the considerable demand for land (for dwellings, industry, and infrastructure) caused by the 10 percent addition to the population of the country as a result of large-scale immigration in the late 1980s and by the expectation that the population would continue to increase at a similar rate (for a more complete discussion of Resolution 533, see the next section).

An additional factor in the weakening of the separation between sectors in the land market is the transfer of the Israel Lands Authority from the Ministry of Agriculture to the Ministry of Construction. The tendency of the Authority is to put commercial considerations and the consideration of lowering housing prices at the center of its policy with the aim of maximizing its contribution to the public purse (as opposed to the policies in earlier years, when settlement and agricultural considerations enjoyed top priority).

The weakening of the separation between sectors in the land market has facilitated the transformation of agricultural land into land for setting up dwellings, industries, and tourism-related installations. This is likely to make it easier for settlements and settlement members to make the most of their major source of wealth, namely, most of the land reserves of Israel, not only at the periphery of the country but also in the central and coastal regions, where it is most expensive.

Some surmise that one of the motives leading to the purposeful weakening of the separation between the settlement and the urban sectors in the land market was to help the settlement sector overcome its financial distress. It is by no means certain, however, that the sector will gain from any changes in zoning, or that the profits after payment of taxes and additional rent will be as large as some of the settlement leadership expects. Nor is it certain that in the case of a partnership between a settlement and private entrepreneurs it is the settlement that will receive most of the benefit of land use change.

In sum, the transition from slow decline until 1986 in the size of agriculture and the employment it provides to rapid decline in 1987 to 1991 may indicate that the partial barriers that separated the settlement sector from the rest of the country mitigated the influence of the general market until 1986 and then became weakened, especially in the capital and land markets, due to the financial crisis of the settlements and the end of their unconditional insurance provided by the state. When barriers weaken, they lose their capacity to moderate the influence of the environment. It is possible that the slowing down of change in 1992 and 1993 indicates that a new equilibrium has been reached, at least for the moment.

## LOOKING TO THE FUTURE

The trend of decooperativization of the moshav may have spent itself simply because there is not much left that is cooperative. Some of the kibbutzim are already embracing further developments, such as unequal salaries for unequal work.

It is not clear whether the situation in the rural sector will be controllable in the future, more so because in the past, the sector whose development was planned by the Settlement Department of the Jewish Agency did not develop tools that might enable it to control its own planning and evolution and because such tools will be now harder to come by because of the processes of decooperativization and decollectivization that have already occurred.

Furthermore, Resolution 533 and Resolution 611 (which replaced it) were but the beginning of the story. While these lines were being written, a government commission (the Fogel Commission) headed by the Director General of the Treasury and in which the ministers of Agriculture and Construction participated came up with new proposals. As a result, Resolution 666 was adopted by the Council of the Lands of Israel. From now on, decisions about where and how to develop will remain the exclusive prerogative of the Council. However, if land is taken for building purposes from the kibbutzim and the moshavim, they will receive substantial compensation (to the tune of 28% of the new value of the land). In the center of the country and in the coastal zone, this may amount to millions of dollars. The way is now open for massive housing construction in the coastal zone and in the center of the country on land formerly defined as agricultural. It is not yet known how this new turn of events will affect the rural sector in general, and the fabric of the moshavim in particular.

One complicating element is a recent public statement from the head of the Prime Minister's office, who, taking notice that Israel (with the exception of the Negev) is more densely populated that any Western country, has said that from now on no single-story buildings will be allowed in the center of the country. It is not clear whether this new policy line will be implemented, and if so, whether it includes the harhavot program in the moshavim. If both answers are positive, the new policy will put the brakes on the gentrification of the moshavim. If the new policy is implemented without affecting the harhavot plans, this will no doubt give a strong boost to the gentrification of the moshavim in the center of the country.

## REFERENCES

Applebaum, L. 1986. "The Migration of Urban Families to Israel's Moshavim 1958–1978." Ph.D. diss. Jerusalem: Hebrew University of Jerusalem.

Applebaum, L., and F. Kedar. 1992. The Expansion Program of Moshav Communities (A Follow-Up Study). Research Report. Rehovot: Development Study Center.

———. 1993. The Expansion Program of Moshav Communities (A Follow-Up Study, Phase 2, 1992). Research Report. Rehovot: Development Study Center (Hebrew).

Central Bureau of Statistics. 1993. *Statistical Annual.* Jerusalem: Central Bureau of Statistics.

Council of the Lands of Israel. 1992. Resolution concerning changes of purpose of agricultural land (Resolution 533). Jerusalem, May 11.

———. 1993. Declaration concerning changes of purpose of agricultural land (Resolution 611). Jerusalem, October 13.

Fleischer, A., T. Banin, and P. Kahana. 1991. *Room and Cottage Renting on the Farm in the Galilee: Survey of the Present Situation and Evaluation of the Potential.* Rehovot: Settlement Studies Center.

Government Comptroller. 1988. Annual Report for the Budget Year 1987. Jerusalem: Government Printing Office (Hebrew).

Ministry of Agriculture, the Authority of Planning and Development of Agriculture, Settlement, and Village. 1994. Annual Economic Report on Agriculture and Village Life for the Year 1994. Jerusalem.

Schwartz, M. 1986. Unemployment Benefits to Moshav Youth in the Galilee. Rehovot: Settlement Studies Center (Hebrew) (photocopy).

Schwartz, M. 1995. *Unlimited Guarantees: History and Political Economy in Israel's Cooperative Agriculture.* Beer Sheva: Ben-Gurion University Press (Hebrew).

Schwartz, M., and L. Applebaum. 1994. *Regional Councils in Times of Change: The Challenge of Development Planning.* Rehovot: Development Study Center (Hebrew).

Schwartz, M., R. Bar-El, A. Nesher, and R. Finkel. 1987. "Moshav-based Industry in Israel." In *Rural Industrialization in Israel,* R. Bar-El, ed., pp. 57–104. Boulder, Colo.: Westview Press.

Schwartz, M., and D. Giladi. 1993. "25% Farmers? Plans and Reality in the First Decade of the State of Israel." *Riv'on Lekalkala (Economic Quarterly)* November: 391–414 (Hebrew).

Sharkansky, I. 1988. *The Political Economy of Israel.* New Brunswick, N.J.: Transaction Books.

Sherman, N., and T. Banin. 1992. *The Attitudes of Moshav Members Regarding the National Moshav Movements: Findings of a Public Opinion Survey.* Rehovot: Development Study Center (Hebrew).

Sherman, N., F. Kedar, and M. Kedmon. 1993. *Non-Agricultural Business in Moshav-Type Agricultural Settlements.* Rehovot: Development Study Center (Hebrew).

Sherman, N., Y. Laks, and F. Kedar. 1994. *Infrastructure for Non-Agricultural Businesses in Areas of Moshav-Type, Planned Agricultural Settlement.* Rehovot: Development Study Center (Hebrew).

Shklanewicz, S. 1983. *Agricultural Structure in Moshav Smallholdings in 1981: Report on the Execution of a Research Plan.* Tel Aviv: Ministry of Agriculture (Hebrew).

Vilkansky, R. 1980. "Kernel and Periphery in the Development of Israel." Dissertation in partial completion of the requirements for the Doctor of Science Degree. Presented to the Senate of the Technion, Technological Institute of Israel, Haifa (Hebrew).

Zussman, P., Y. Kislev, Z. Lerman, Y. Kroll, and G. Rosenthal. 1990. "The Credit Crisis in Agriculture." *Economic Quarterly* 143: 403–437 (Hebrew).

Chapter 14

# Rural Cooperation and the Stages of Economic Growth

*Raanan Weitz*

*In this final chapter, Raanan Weitz draws upon his unparalleled experience as a leader in regional planning in Israel to provide an evolutionary overview of how the relationship between government support and private enterprise in rural cooperative organizations changes through time with economic growth. Earlier chapters examined this change as it appeared from the perspective of individuals and individual communities; later chapters were concerned with what might be regarded as failures or terminations or, alternatively, as transitions in the state of Israeli rural cooperatives. Weitz suggests both a way of evaluating the appropriateness of the balance between government and private input and a way of anticipating future change.*

This chapter presents several theses on the relations between the stages of economic growth and the organizational characteristics or rural cooperation, specifically with regard to agriculture and related activities. We will begin with a formal description of the two systems in question: the system of growth in the general and agricultural economy and the system of rural cooperation.

## THE PROCESS OF ECONOMIC GROWTH

Economic growth both influences and is influenced by social changes. This relationship pertains even during periods of stagnation or negative growth. For purposes of this study, we have used average GNP (in 1988 dollars) as the yardstick for measuring economic growth. Although the process is a continuous one, we have divided it into three stages in accordance with the relationship between demand for goods and services and the nature of the output of supply. Table 14.1 presents the main characteristics of the growth process in each of its three stages.

The term *physical infrastructure* refers to the distribution of road, rail and air transport, communications, energy, water drainage, and other systems throughout all regions of a country. Income in rural regions has been determined on the basis of the averages in nonmetropolitan regions compared with metropolitan averages based on 1988 dollar values (Weitz 1986).

In this section, I deal only with the three dynamic stages of growth, omitting the traditional pregrowth stage. This is because

**Table 14.1**
**The Stages of Economic Growth: Main Characteristics**

| | | Economic Growth Process | | |
|---|---|---|---|---|
| | Pregrowth traditional economy | Stage one: Basic needs | Stage two: Social competition | Stage three: Affluence |
| Relationship between Output and Demand | Static balance | Output fails to meet demand | Output meets demand | Output meets all possible demand |
| GNP per person (1988$) | 200–300 | 500–2,000 | 3,000–20,000 | Over 30,000 |
| Employment structure | Mainly agriculture | Balance between sectors and services | Mainly industry | Mainly services. Industry is white collar |
| Physical infrastructures | Insignificant | Uneven distrib. Large regional diffs. | Full distrib. Small reg. diffs. | Complete distribution |
| Rural income | Low | Significant growth | Approaching urban income | No diff. between rural and urban income |
| Rural urban relations | Society is mainly rural | Growth of rural towns | Growth of regional cities | Rural-urban differences disappear |

during the traditional stage, cooperative frameworks as we define them did not exist and because the traditional economy is in the process of disappearing permanently from the world scene. To be sure, the traditional economy still prevails in several regions (Africa, parts of China, India, and parts of Latin America). But these are passing phenomena. There also exist islands of underdevelopment in developed countries (i.e., the Amish sect in the United States) which intentionally preserve a lifestyle based on a traditional economy. However, these cases are the exception that prove the rule. In fact, most countries today are in the first stage of growth, a minority are in the second stage, and a few are on the threshold of the third stage.

The traditional stage has prevailed in the formative period of every human society under varying geographic conditions and with varying levels of access to natural resources. In each case, however, two basic developmental characteristics have been present: The first is a static tendency, lasting for long periods of time (i.e., China, Egypt, Babylon), whereby the population grew in accordance with the level of food output. The second characteristic is the percentage of the people engaged in nonagricultural pursuits as a direct function of the capacity of the average agricultural family to generate food surpluses. The nonagricultural sector represented between 10 and 15 percent of the population of the traditional economy. Throughout the entire traditional period, the growth multiplier was almost zero.

Economic growth emerged with the ability of the average agricultural family to produce ever growing quantities of food. This began in those countries where the first industrial revolution took place. The economic system became increasingly dynamic as production technologies in all three sectors underwent continuous improvements and modifications.

Economic growth is the result of ever increasing production output per unit of human labor. This increase is the result of two basic factors: improved productive technologies and organizational changes in the use of these technologies. These factors depend on the level of investment in improving the means of production and the utilization of these improvements by social systems impinging on organizational frameworks. These two elements—investment and organizational change—can derive either from public or private sources. For our purposes, the term *public source* refers to those powerful enough to impose their will on others. Formulated in another way, there are two engines of economic growth: the government and its agencies and private enterprise in its various forms. These two forces promote growth and the distribution of its fruits.

The rate of growth is determined by the behavior of government and individuals. The ratio between these two forces, which creates a maximum growth rate under conditions of available resources, we call a *mix*. This mix expresses a relationship between financial means and organizational frameworks, which are used by each of the two forces in its own area of economic activity and in a specific region. For purposes of the current presentation, I make the following assumptions regarding the behavior of these two forces: (1) In every human society, there exists for every situation a specific mix of the two forces that creates the maximum possible rate of growth in the light of existing economic, natural, and human resources. I call this mix the *optimal mix.* (2) The term *every situation* is defined here by two dimensions: the *time dimension,* which is determined by the level (not the rate) of economic growth, and the *space dimension,* which is determined by the territorial level of government decision making.

The optimal mix is a dynamic concept of changing values in different situations, defined by the above two dimensions. Since natural resources are given in every situation, the optimal mix depends only on the relationship of the two motive forces of economic growth. The availability of natural resources may assist in achieving a faster rate of growth and, as a result, faster changes in the optimal mix, but it cannot effect the mix itself. This explains why poor nations sometimes exist in rich countries and vice versa. It also explains regional differences within the same country.

In light of these assumptions, we will attempt to draw practical applications for the various stages of economic growth. Table 14.2 describes the proposal. The table is built around hypothetical situations based on three levels of economic growth, concurrent to the stages outlined in Table 14.1, as well as three levels of government organization: the national level, the regional level, and the local level of production units. It should be emphasized that we have omitted the federative structure existing in large countries (i.e., the United States, China, Russia, India, Brazil) in order to simplify the proposal.

## GROWTH AND THE STRUCTURE OF THE FARM ECONOMY

Research indicates that there exists a certain functional connection between the structure of the agricultural production unit and the level of economic growth (Weitz 1979). The structure of the production unit in the farm economy has a decisive influence on the socioeconomic function of the rural community, particularly in

**Table 14.2**
**The Optimal Mix of Development Investment: Government Investment as a Percentage of Total Investment in Development**

| | | Stage 1: Basic Needs | Stage 2: Social Competition | Stage 3: Affluence |
|---|---|---|---|---|
| National level | Of total investment in the country | 30 | 20 | 15 |
| | Share in national enterprises | 10 | 35 | 40 |
| | Share in regional enterprises | 50 | 35 | 40 |
| | Share in production units | 40 | 30 | 20 |
| Regional level | Share of total investment | 60 | 30 | 25 |
| Local level | Agriculture | 35 | 15 | 5 |
| | Industry | 40 | 20 | 10 |
| | Economic services | 100 | 80 | 60 |
| | Social services | 70 | 50 | 40 |
| | Commercial services | 10 | 0 | 0 |
| | Public Admin. services | 100 | 100 | 90 |

From Raanan Weitz, "The Optimal Mix Principle," in *From Peasant to Farmer* (New York: Columbia University Press, 1971), pp. 109–132.

the transition from the traditional economy (which produces almost exclusively for personal consumption) to the modern economy (which focuses on production to satisfy market demand). Table 14.2 does not include the structure of the traditional farm economy that produced for the direct consumption of the farmer's family.

The findings presented in Table 14.3 raise two questions: First, is there any need for the mixed economy stage? Is it not possible to move directly from traditional subsistence agriculture to specialized farming? Second, were not the great plantation estates established

**Table 14.3**
**Stages of Growth and the Structure of the Farm Unit (Figures Based on 1988 Dollar Values)**

| | Stage 1 Basic Needs | Stage 2 Social Competitions | Stage 3 Affluence |
|---|---|---|---|
| GNP/person | 500–2,000 | 3,000–20,000 | Over 30,000 |
| Nature of farm economy | Diversified | Specialized | Super-specialized |
| Composition of the farm | Varied | Main Branch | Single branch |
| Yearly work schedule | Balanced | According to farm type | Mainly seasonal |
| Destination of product | Direct marketing and personal consumption | Direct marketing and processing | Processing |
| Value of output | 1,500–6,000 | 8,000–40,000 | Over 100,000 |
| Ratio added value to output | 40–50 | 25–35 | Less than 20 |
| % inputs from non-agricultural sector | 10–15 | 20–30 | Over 40 |

by the eximperialist powers in their colonies really specialized farms, something which would make the transition easier?

The answer to the first question is negative for the following reasons: The diversified economy is labor-intensive and aims at maximum utilization of the work potential of the entire farming population. This structure is best suited to the gradual accumulation of capital required to make the transition to the specialized farm economy, which is fundamentally capital-intensive. The transition from a mixed to a specialized economy takes place gradually but ultimately encompasses all agricultural regions in the country. This transition required tremendous investments in physical infrastructures, education, and vocational training for the large numbers of farmers, particularly in nonagricultural sectors, in order to absorb the manpower that has left the agricultural work force without lowering the rate of growth in the overall purchasing power of the population as a whole.

The answer to the second question is also negative. Although the colonial era plantations did specialize in one crop, they in fact constituted a camouflaged form of the traditional farm economy, since their existence depended on the exploitation of large amounts of cheap labor at subsistence wages. The plantations are labor-intensive farm units, unable to accumulate the capital necessary for economic growth or for expanding the purchasing power of the masses.

## WHAT IS COOPERATION?

The cooperative idea is relatively old. It began during the last century with Peter Kropotkin's proposal to establish farms administered by common agreement of the workers, managers, and workers. Over the years, the idea developed and deepened. Fundamentally, the cooperative idea is an attempt to utilize cooperative relationships between individuals as a motivating factor in economic progress and improving the quality of life. Thus, for example, cooperation began in Palestine at the beginning of this century as an attempt by owners of citrus orchards and vineyards to improve their own economic situation by means of cooperative marketing organizations. Following this, other cooperatives were established by small farmers for the purchase of farm inputs and in order to secure credit for operating capital. The cooperative movement rose primarily as an effort by farmers to confront the network of intermediaries, which at the time was an integral part of the traditional system that prevailed in rural areas.

As this simple concept proliferated along with attempts to implement it in practice, the cooperative idea encompassed an increasingly wide range of organizational forms: from the kibbutz, which is characterized by full cooperation, both in economic and day-to-day life, to the various collective organizations wherein cooperation is limited to the economic sphere, leaving other areas of activity to the family (the kolhoz, commune, cooperative moshav, and other functional cooperatives); from communities based on family farms, united in cooperative frameworks to provide services and administrative activities, to cooperatives which are not identified with their members' places of residence and that deal only with specific areas of business activity.

The purpose of a cooperative is to carry out defined tasks in the field of production and services. The management of the cooperative is chosen for a fixed period of time. The tasks of the cooperative, its organizational arrangement, and the selection of its management may be determined either by the free will of its members or by government fiat. Property in the cooperative is owned jointly by all its members to an equal degree. This last characteristic is the primary difference between cooperation and other forms of organization. Equal ownership of property determines the distribution of the cooperative's profits but not the rate of salaries within the cooperative.

## COMPULSORY AND VOLUNTARY COOPERATIVES

In compulsory cooperatives, membership, selection of management, and the setting of objectives is determined by government. Therefore, they are essentially government cooperatives. In voluntary cooperatives, the above-mentioned aspects are determined by the free will of the members of the cooperative. They are therefore private cooperatives and operate on the basis of private initiative. Accordingly, the efficiency of a government cooperative will be measured by its ability to achieve the development objectives set by the government. By contrast, the efficiency of private cooperatives will be measured by the profits generated by the activities that they are engaged in.

Cooperatives can be divided into three types in accordance with the nature of the activity which they were established to carry out.

1. *Production cooperatives.* The present study will deal only with those engaged in agriculture.
2. *Economic services cooperatives.* Their purpose is to provide economic services exclusively. Our concern is with the supply of agricultural production services, such as marketing, farm equipment, credit, agricultural vocational training, professional guidance, and the like. Service cooperatives also establish and manage industries that are directly linked to agriculture.
3. *Social–municipal service cooperatives.* Their purpose is to provide social-municipal services, such as education, health, the supply of household goods, culture, sanitation, transportation, communications, and so forth.

There are also *multipurpose cooperatives* that deal with more than one of the fields listed above. The moshav, for instance, unites economic and municipal services in the framework of a single cooperative. (To the best of my knowledge, the moshav is unique in this respect.) In each of the three fields, a cooperative may be unibranch (in other words, deal with only one branch of activity) or multibranch (dealing with more than one branch in the same field). In this respect, the moshav cooperative society is multibranch and multipurpose.

## THE STRUCTURE OF RURAL COOPERATIVES AND THE STAGES OF ECONOMIC GROWTH

Based on the experience that has been accumulated over the past fifty years as to the function of agricultural cooperatives in national and regional economic systems, as well as with regard to the production units themselves, we can reach a number of conclusions. The conclusions outlined in the following paragraphs are based on the personal experience and research of the author (Weitz 1971).

First, all production cooperatives, both compulsory and voluntary, that were set up to handle the agricultural production process itself proved inefficient and ultimately failed. All attempts to create a large, administered farm instead of the family farm inevitably failed as well. This applies to all the collective farms (the kolhoz in Russia, the commune in China and Cuba, the ejido production cooperatives in Mexico, and the like); to the large, government-owned and administered farms (the sovhoz in Russia, the government farms in Romania, Poland, and Egypt); and to the large private estates in Europe, Latin America, and India.

These conclusions are based on an analysis of the function of the cooperative in agricultural production. Like industry, agriculture involves activities that benefit from economies of scale. Unlike industry, however, activities that require the immediate involvement of producers with their products play a very significant economic role. This is because the biological process does not allow for complete standardization of the production process. The vast changes that agrotechnology has undergone as a result of innovations in genetics and microbiology have not changed this fact. For this reason, the family farm has been able to compete with large, managed farms, both collective and private, compulsory or voluntary.

When it comes to activities that require economies of scale, the success of the family farm depends on support systems (Table 14.3). It is here that the economic services cooperative plays a decisive role. But it can play this role only if both the structure of the family farm and of the cooperative are flexible and adaptable to technological changes in the production process.

The kibbutz is the only community framework that defies this rule by combining economies of scale with direct, family attention. This will hold true as long as the kibbutz continues to function like a large family.

As the farm economy makes the transition to the intensive, super-specialized production phase, it is quite possible that the large, managed farm will be able to operate efficiently and profitably by employing computerized, automated facilities that could overcome the distinction between activities that require direct attention and those that require economies of scale.

Second, national and regional production cooperatives do not exist, and I have come across no proposals to establish them. The closest example of a regional production cooperative is the group of cooperatives in Mondragon, Spain. But Mondragon is not a single, multibranch agricultural cooperative but rather a group of multipurpose, nonagricultural cooperatives that exist in one city.

Regional cooperatives did at one time (and some still do) exist in Israel. The reference here is to the regional organizations, which consisted of service cooperatives belonging to rural communities rather than private members. Although these were fundamentally service cooperatives, they also set up and managed factories for processing agricultural produce (i.e., processing plants, slaughterhouses, storehouses for animal feed). In many cases, they even established industrial enterprises having nothing to do with agriculture or managed centralized agricultural branches (cotton fields, citrus orchards, etc.), thus overstepping their objectives and the boundaries of their competence.

Third, the combination of service and social-municipal cooperatives in one framework can function only during the first stage of economic growth. With the transition to the specialized farm, this combination becomes inefficient and even counterproductive. Specialization reduces the number of farmers. The ability to control the migration to the cities and to preserve the viability of rural communities depends on the generation of local, nonagricultural employment within the rural regions themselves. At this stage, it is necessary to maintain an organizational separation between economic services and social-municipal services.

Fourth, when the structure of the farm unit reaches the specialization phase, each production unit requires ever-increasing and more specialized economic support services (marketing, supply, credit, guidance, and so forth). At the stage when several specialized farms focusing on different branches exist within one village, a regionwide specialized single-branch business organization is called for to service the various specialized farms. The voluntary, regional, single-branch service cooperative can be effective in this regard. The determining factor in the establishment of such a cooperative is its cost-effectiveness for its member production units. Such an organization can also be a holding company, or even a private concern, rather than a cooperative. Experience in European countries has shown that voluntary cooperatives of this type are economically competitive with private organizations and more effectively promote the interests of their members.

It is possible that with the advent of super-specialization the regional organization will once again prove inadequate. Services will then be provided by nationwide, specialized business organizations with the capacity to deal directly with all the production units in a particular branch throughout the country. This will be made possible by future technological advances in all service systems (transport, communications, information, banking), as well as by the advantages of economies of scale that a nationwide organization can provide.

Fifth, the transition from a diversified to specialized and super-specialized structures requires that municipal matters be dealt with on a regional level rather than on the level of an individual community. Since rural villages are becoming less agricultural in nature as nonagricultural employment proliferates, their social-municipal needs increasingly approximate those towns or cities. It will therefore be necessary to establish a regional framework to service both the farm population and other residents of rural villages. This raises the possibility of establishing integrated regional municipalities to deal with social-municipal matters, something that is likely to prove most effective.

Sixth, all cooperative organizations require a system of constant government oversight. This must be carried out by a special unit established with the objectives of the various cooperatives in mind and appropriate to evolving social conditions. Government authorities must therefore establish a special mechanism for this purpose. Table 14.4 details the nature of the cooperative system throughout the various stages of growth, as previously described.

The world views that have prevailed since the end of the last century have become outdated and will definitely prove unsatisfac-

**Table 14.4**
**Cooperation and the Economic Growth Process (Proposed Model)**

| | Stage 1<br>Basic Needs | Stage 2<br>Social Competitions | Stage 3<br>Affluence |
|---|---|---|---|
| Structure of farm economy | Diversified | Specialized | Super-specialized |
| Community level | Voluntary multi-purpose economic and socio-municipal coops. | Individual farms directly connected to specialized regional coops. | Individual farms served directly by vertical specialized national cooperatives |
| Regional level | Multi-purpose regional service cooperatives | Specialized regional cooperative | Regional urban-rural municipality |
| National level | Government unit for oversight of multi-purpose village service coops.<br>Government unit for oversight of regional cooperatives | Government unit for oversight of regional economic cooperatives | Specialized cooperatives directly connected to individual producers |

tory in the future. Neither classic capitalism, which saw private enterprise as a universal solution, nor Marxism-Leninism, which saw a panacea in government action, can explain the development processes unfolding in our generation. Rapid advances in production technologies and organizational methods require that we change the way in which we look at economic development. The thesis presented herein is based on the concept of the optimal mix and on the possibility of separating employment mobility from geographical mobility. It attempts to outline a new perspective on the forces that motivate the economic growth process, taking into account social and personal values in order to improve the welfare of individuals, their families, and their communities.

In light of recent changes in the communist world, the following observations should be made regarding the People's Republic of China and the former Soviet republics. The Chinese economy is still at the first stage of development (basic needs). Therefore, the Chinese model of development should follow the recommendations outlined in first column of Table 14.4, namely, "basic needs." The former Soviet republics, on the other hand, are at the second stage (social competition). Therefore, their model of development should follow the recommendations outlined in the second column, namely, "specialized" structures. One key point, however, is common to both instances. Both the Chinese and Soviet farm economies must make a basic transition from large, collective farms to the family farm unit.

The cooperative as described herein can help in achieving this. Like all other human institutions, the cooperative must adapt its structure and activities in light of economic growth.

## REFERENCES

Weitz, R. 1971. *From Peasant to Farmer*. New York: Columbia University Press.

———. 1979. *Integrated Rural Development*. Rehovot: Settlement Studies Center.

———. 1986. *New Roads to Development*. Westport, Conn.: Greenwood Press.

# Index

# About the Contributors

**Shlomo Fox** was born in Manchester, England, in 1945 and has economics degrees from the Universities of Oxford and Sussex. In 1970, he moved to Israel to Moshav Noam in the northern Negev where he immersed himself and his family until 1972, taking a rare opportunity to observe the political economy of an immigrant village and its interface from the clients' angle with the development bureaucracy. Since then, he has worked in that bureaucracy at the Jewish Agency in the field of rural development, and most recently, within the newly established Department for Rural and Urban Development. He has been very occupied in the area of small business loan funds and technological incubators in the Galil and Negev.

**Harvey E. Goldberg** is a social anthropologist in the Department of Sociology and Anthropology of the Hebrew University of Jerusalem. His research deals with ethnicity in Israel and with the social history of North African Jewish communities. His most recent book is *Jewish Life in Muslim Libya: Rivals and Relatives* (Chicago University Press, 1990).

**Gideon M. Kressel** is head of the Social Studies Center at the J. Blaustein Institute, Ben-Gurion University of the Negev, Israel. He is currently studying patterns of cooperation and competition at work and in the marketplace. His field work is centered upon the Bedouin market of Beer Sheva and the open-air markets in Varna, Bulgaria, the kibbutz in transition away from egalitarianism, and the regression of the cooperative set up in the moshav.

**Susan Lees** (Ph.D., University of Michigan, 1970) has been a faculty member in the Department of Anthropology at Hunter College, City University of New York since 1971. A year before that, she taught in the Department of Ancient Middle East Studies at Tel Aviv University. Her initial area of ethnographic research was in Latin America; but while working with Israeli technical experts there, she became fascinated with differences in the Israeli approach to rural development from that of other nations. In 1983, she was privileged to renew her acquaintance with these experts in Israel and used the opportunity to initiate a study of irrigated agriculture in Israel. For eighteen years (1977–1995), she served as editor of an interdisciplinary journal, *Human Ecology.* She has written extensively about irrigation and society in Israel, Latin America, and elsewhere and about the relationship between bureaucracy, power, and resource management.

**Yair Levi** was born in Italy and emigrated to Israel. He received his B.A. and M.A. degrees from the Hebrew University of Jerusalem and his Ph.D. in political sciences from the State University of Liege, Belgium. He is Scientific Director of CIRCOM (International Research Center on Rural Cooperative Communities) and editor of the *Journal of Rural Cooperation.* He has been senior researcher at the Development Study Center, Rehovot and senior lecturer at Haifa University. He has published several articles on cooperatives and community and is author of *Realta e Utopia in una Comunita Siciliana,* coeditor (with H. Litwin) of *Communities and Cooperatives in Participatory Development,* and coauthor (with G. Naveh) of *Toward Self-Management in New Land Settlement Projects.*

**Leonard Mars** is a lecturer in Social Anthropology at the University of Wales, Swansea. He studied anthropology at the universities of Edinburgh and Manchester from where he obtained his doctorate for research on the moshav Givah, which was published under the title *The Village and the State* (1980). He has conducted research in Newfoundland, Israel, South Wales, and Hungary. He is the author of various papers on the anthropology of Jews and Judaism. His most recent publications include "Coming of Age among Jews: Bar Mitzvah and Bat Mitzvah Ceremonies" in *Anthropology and the Riddle of the Sphinx: Paradoxes of Change in the Life Course,* Paul Spencer, ed. (Routledge, 1990); "The Ministry of the Reverend Simon Fyne in Swansea: 1899–1906" in *Jewish Social Studies*; "The Incorporation of a Stranger: Analysis of a Social Situation in a Welsh Valley" in *The Jewish Journal of Sociology.*

**Naomi Nevo** emigrated in 1946 from England to Palestine. She did field work in kibbutzim toward her M.A. degree concomitant with a position as lecturer on Social and Political Institutions at the British Council School of Higher Studies. After serving in the Israeli War of Independence and a brief period in New York at the United Nations, she returned to Israel in 1954 as Research Assistant to Minister of Defense, David Ben-Gurion. From 1959 to 1986, she served as sociologist in various regions of the Jewish Agency Rural Settlement Department, including the last two years as director of the Department's team of sociologists and anthropologists. She is currently engaged in ongoing research on veteran rural communities, position and role of rural women, and changing cooperative frameworks.

**Esther Schely-Newman** is a lecturer in the Noah Mozes Department of Communication and Journalism at the Hebrew University of Jerusalem. Her current research treats the oral history of Israelis who immigrated from Afro-Asian countries, particularly settlers in moshavim. Her special interest is narrative modes as indications of retention of ethnic identity.

**Moshe Schwartz** is Senior Researcher at the Development Studies Center at Rehovot and also doing research at the Social Studies Center at the J. Blaustein Institute, Ben-Gurion University of the Negev, Israel. He is currently studying agricultural cooperatives in Africa and Nepal. Some of the subjects he has studied are the transformations of the Israeli Moshav, regional organization, the breakdown of cooperative organizations, and agricultural research and development organizations in Israel. His most recent book is *Unlimited Guarantees: History and Political Economy in Israel's Cooperative Agriculture* (1995), published by Ben-Gurion University Press (in Hebrew).

**Neal Sherman** is Senior Researcher at the Development Study Center, Rehovot, Israel. He also serves as coordinator of a special training and consultancy project aimed at moshavim that are establishing new residential neighborhoods. Sherman has published and lectured on the moshav and on rural development policy and administration. He is coauthor of *Cooperatives and Development: Agricultural Politics in Ghana and Uganda* (1982) with M. C. Young and T. Rose.

**Moshe Shokeid** (Ph.D., Manchester, 1968), Professor of Anthropology at Tel Aviv University, has studied minorities: North African immigrants, Israeli Arabs, and Israeli emigrants in New York. Recently, he studied an American gay synagogue. Among his publications are *Children of Circumstances* (1988), *Distant Relations* (1974) with S. Deshen, *The Dual Heritage* (1971, 1985), and *A Safe Space* (forthcoming).

**David Sivan** was employed by the Ministry of Agriculture (Negev region) in the Division of Planning during 1984/85. By the end of 1985, he joined the same division of the Settlement Department of the Jewish Agency. During the whole period, he was involved in an administrative position to help channel the government's funds in the context of the Law of Encouraging Capital Investment in Agriculture to the moshavim of the Negev. He also has been involved with financial problems in these moshavim. In this context he was involved with the development program of Moshav Ranen, founded in 1988. In particular, Moshav Ranen engaged in a successful investment program in greenhouse agriculture when other moshavim experienced financial crises and economic decline. Against this background, he initiated the current research to find out the reasons for the unusual and successful behavior of this particular community. In addition, he had conducted several studies concerning economic feasibility of new ideas connected with agricultural research and development.

**Raanan Weitz** is head of the Development Study Center, Rehovot, Israel. For many years, he was actively responsible for establishing agricultural settlements in Israel for new immigrants coming from Europe, the Middle East, and North Africa. He also has acted as a consultant to many international agencies and taken part in rural development programs throughout the developing world. Since the early 1950s, he has researched moshavim in Israel. In 1991, he received the Israel Prize for his contribution to the development of research about agriculture and regional planning.

ISBN 0-275-95309-2

90000>

EAN

9 780275 953096

HARDCOVER BAR CODE